T0210696

Logic Locking

Dominik Sisejkovic • Rainer Leupers

Logic Locking

A Practical Approach to Secure Hardware

 Springer

Dominik Sisejkovic
RWTH Aachen University
Aachen, Germany

Rainer Leupers
RWTH Aachen University
Aachen, Germany

ISBN 978-3-031-19122-0   ISBN 978-3-031-19123-7   (eBook)
https://doi.org/10.1007/978-3-031-19123-7

© The Editor(s) (if applicable) and The Author(s), under exclusive license to Springer Nature Switzerland AG 2023

This work is subject to copyright. All rights are solely and exclusively licensed by the Publisher, whether the whole or part of the material is concerned, specifically the rights of translation, reprinting, reuse of illustrations, recitation, broadcasting, reproduction on microfilms or in any other physical way, and transmission or information storage and retrieval, electronic adaptation, computer software, or by similar or dissimilar methodology now known or hereafter developed.

The use of general descriptive names, registered names, trademarks, service marks, etc. in this publication does not imply, even in the absence of a specific statement, that such names are exempt from the relevant protective laws and regulations and therefore free for general use.

The publisher, the authors, and the editors are safe to assume that the advice and information in this book are believed to be true and accurate at the date of publication. Neither the publisher nor the authors or the editors give a warranty, expressed or implied, with respect to the material contained herein or for any errors or omissions that may have been made. The publisher remains neutral with regard to jurisdictional claims in published maps and institutional affiliations.

This Springer imprint is published by the registered company Springer Nature Switzerland AG
The registered company address is: Gewerbestrasse 11, 6330 Cham, Switzerland

Preface

A subtle change that leads to disastrous consequences—hardware Trojans undoubtedly pose one of the greatest security threats to the modern age. How to protect hardware against these malicious modifications? One potential solution hides within logic locking, a prominent hardware obfuscation technique. In this book, we take a step-by-step approach to understanding logic locking, from its fundamental mechanics, over the implementation in software, down to an in-depth analysis of security properties in the age of machine learning. This book can be used as a reference for beginners and experts alike who wish to dive into the world of logic locking, thereby having a holistic view of the entire infrastructure required to design, evaluate, and deploy modern locking policies.

Aachen, Germany   Dominik Sisejkovic
Aachen, Germany   Rainer Leupers
August 2022

Contents

Acronyms

3PIP	Third-Party Intellectual Property
AES	Advanced Encryption Standard
AGR	AppSAT-Guided Removal
AIG	And-Inverter Graph
ALU	Arithmetic Logic Unit
ANN	Artificial Neural Network
ANT	AND Netlist Test
APD	Area-Power-Delay
ASIC	Application-Specific Integrated Circuit
AST	Abstract Syntax Tree
AT	Area-Timing
ATI	AND-Tree Insertion
ATPG	Automatic Test Pattern Generation
BEOL	Back End of Line
BeSAT	Behavioral SAT
BFS	Breadth-First Search
BISA	Built-In Self-Authentication
C1HT	Class-1 Hardware Trojan
C2HT	Class-2 Hardware Trojan
CAA	Cycle Analysis Attack
CLC	Control-Lock Circuit
CLIC-A	Characterization of Locked Integrated Circuits via ATPG
CMOS	Complementary Metal-Oxide Semiconductor
CNN	Convolutional Neural Network
CPU	Central Processing Unit
D-MUX	Deceptive Multiplexer Logic Locking
DAG	Direct Acyclic Graph
DC	Design Compiler
DDIP	Double DIP
DFA	Differential Fault Analysis
DfTr	Design for Trust

DIP	Distinguishing Input Pattern
DLCL	DIP Learning on CAS-Lock Attack
DoS	Denial of Service
DPA	Differential Power Analysis
eD-MUX	Enhanced D-MUX
EDA	Electronic Design Automation
FEOL	Front End of Line
FLL	Fault Analysis-Based Logic Locking
FPGA	Field-Programmable Gate Array
GA	Genetic Algorithm
gD-MUX	Generalized D-MUX
GE	Gate Equivalent
GNN	Graph Neural Network
GSS	Generalized Set Scenario
HCA	Hill-Climbing Attack
HD	Hamming Distance
HDL	Hardware Description Language
HLS	High-Level Synthesis
HT	Hardware Trojan
HW	Hardware
IC	Integrated Circuit
IFS	Identify Flip Signal
ILC	Inter-Locking Circuit
IO	Input and Output
IP	Intellectual Property
ISA	Instruction Set Architecture
KBM	Key-Bit Mapping
KPA	Key Prediction Accuracy
LC	Layout Camouflaging
LCALL	Logic Cone Analysis Logic Locking
LCBFA	Logic Cone-Based Brute-Force Attack
LCSBLL	Logic Cone Size-Based Logic Locking
LUT	Lookup Table
LVE	Locality Vector Extraction
MiG-V	Made in Germany RISC-V
ML	Machine Learning
MLP	Multi-Layer Perceptron
MUX	Multiplexer
NAS	Neural Architectural Search
OCP	Optical Contactless Probing
OG	Oracle-Guided
OL	Oracle-Less
OOP	Object-Oriented Programming
PF	Point Function
PSA	Path-Sensitization Attack

PSO	Particle Swarm Optimization
RAA	Rationality Analysis Attack
RARLL	Redundancy Attack Resistant Logic Locking
RE	Reverse Engineering
RLL	Random Logic Locking
RNT	Random Netlist Test
RSAS	Robust SAS
RTL	Register-Transfer Level
SAAM	Structural Analysis Attack on MUX-Based Locking
SARO	Scalable Attack-Resistant Obfuscation
SAS	Strong Anti-SAT
SAT	Boolean Satisfiability Problem
SFLL	Stripped-Functionality Logic Locking
SGS	Sensitization-Guided SAT
SKRA	Statistical Key Recovery Attack
SLL	Strong (Secure) Logic Locking
SMT	Satisfiability Modulo Theory
SoC	System-on-Chip
SPS	Signal Probability Skew
SRS	Self-Referencing Scenario
TAAL	Tampering Attack on Any Key-Based Logic Locked Circuit
TAL	Test-Aware Locking
TDM	Test-Data Mining
TGA	Topology-Guided Attack
TGARLL	TGA-Resistant Logic Locking
TRENTOS	Trusted Entity Operating System
TRLL	Truly Random Logic Locking
TTLock	Tenacious and Traceless Logic Locking

Notation

C^f	(C^f) Binary Galois field 2
D^f	Functional deceptiveness factor
FHS	Functional hardware security
G^f	Golden factor
H	Frequency histogram
IC_{ll}	Logic-locked design
IC	Original (unlocked) design
I	A single input pattern
KSS	Key-space size
K	Activation key
O	A single output pattern
SHS_{cc}	Structural complexity change
SHS_{en}	Structural key-gate entropy
SHS	Structural hardware security
S^f	Secrecy factor
TT	Truth table
$\mathcal{I}$	Set of all input patterns
$\mathcal{O}$	Set of all output patterns
d_e	Euclidian distance
f^{dc}	Key-space decrease function
f^{kd}	Key-gate distribution measure function
s_d	Structural distribution vector
s_{max}	Maximal structural distance vector
s_{opt}	Optimal structural distribution vector
v_{ext}	Extracted feature vector
v_{max}	Maximal distance feature vector
v_{opt}	Optimal feature vector

## Part I

Hardware Security and Trust: Threats and Solutions

Chapter 1
Introduction

Computer security has become a driving force in the design of modern electronics systems. Over many years, security primitives, specifically in software, have been extensively researched. Hardware (HW) security, in comparison, is a relatively young field, since HW has been traditionally considered immune to attacks, representing a root of trust for any electronic system. However, over the last three decades, an increasing number of vulnerabilities have been identified with the root cause in the hardware itself [29]. Attacks that exploit these vulnerabilities can be broadly separated into two categories. The first category encompasses all attack vectors that are enabled due to an overlooked construction fault in the HW implementation, opening the door for a range of attack vectors, predominantly in the form of side-channel attacks and exploits of other unintended HW side effects. Notable examples in the last years include transient execution attacks, such as Meltdown [106] and Spectre [93], as well as security exploits in dynamic random-access memories, such as RowHammer [92, 118]. The second category includes more recent attack types that are enabled by intentional, malicious changes in the HW, commonly known as Hardware Trojans (HTs) [30]. The challenges introduced by these modifications have a deep impact on the research and development landscape of hardware security, particularly as they can serve as key enablers of a theoretically unlimited attack surface; including information leakage, reliability degradation, and denial of service, among others.

The introduction of HTs yields an interesting question: what is their root cause? Nowadays, a highly competitive environment, short time-to-market, and the ever-increasing need for reduced design and production costs have transformed the Integrated Circuit (IC) supply chain into a global effort, driven by third-party Intellectual Property (IP), subcontracting external design houses, and outsourcing the fabrication to off-site foundries. This deep reliance on external parties has led to a far-reaching consequence—the loss of trust and assurance. Hence, legitimate IP owners are faced with the possibility of injected HTs, leading to untrustworthy HW components. And this challenge is, by all means, a serious one. A wide range

© The Author(s), under exclusive license to Springer Nature Switzerland AG 2023
D. Sisejkovic, R. Leupers, *Logic Locking*,
https://doi.org/10.1007/978-3-031-19123-7_1

of engineers are involved in the design and production of HW, thereby having full access to a design and often operating across multiple organizations, countries, and even continents. It only takes a single rogue entity, implanting a tiny, stealthy, and carefully placed modification, to lay the foundation for a catastrophic attack. This covert nature of HTs makes it difficult to catch them in the wild, in particular, due to the inherent complexity of modern circuits and shrinking feature sizes. Thus, only a handful of alleged HTs have been reported. For example, more than a decade ago, a Syrian radar system failed to warn of an incoming airstrike, reportedly because of HTs embedded in the defense systems [2, 114]. Even though it is difficult to verify the inclusion of HTs in such incidences, the very potential of this tiny, malicious design modification has become a focal point within research and industry. The US military and intelligence executives have placed hardware Trojans among the most severe threats the nation might face in the event of war [111]. Moreover, the US Defense Advanced Research Project Agency (DARPA) has issued multiple funding programs to address the issue of trustworthy electronics, including the TRUST [48], IRIS [46], and SHIELD [47] program, among others. The seriousness of this issue has also been recognized within Germany. The German Federal Ministry of Education and Research (BMBF) has issued a framework program for 2021-2024 to tackle the challenges of trustworthy and sustainable microelectronics for Germany and Europe [32] with a range of projects already underway [33].

The efforts in mitigating HTs evolve around two focal points: detection and prevention. Trojan detection aims at detecting and removing potential HTs, possibly before these are placed in silicon. However, detection approaches are still far from a complete solution due to multiple reasons. First, HTs can be injected on many different levels of the HW design abstraction and in various stages of the IC supply chain. This makes it challenging to derive an effective detection mechanism. Second, even if comprehensive (and often destructive) reverse engineering procedures are deployed to verify the absence of Trojans in chips after production, this does not guarantee that all produced ICs are HT-free. Therefore, more focus has been given to preventing HT insertion by design. In particular, logic locking has evolved as a premier technique to protect against HT insertion by means of key-controlled functional and structural design changes that aim at protecting the asset—the HW design—throughout the IC supply chain [218]. Hereby, the defensive mechanism is built on the assumption that an attacker is required to perform extensive reverse engineering to insert and construct an intelligible, design-specific hardware Trojan. Hence, the locking-induced changes increase the complexity of the attack by binding the correct behavioral and structural HW characteristics to a secret key. Nevertheless, the evolutionary timeline of logic locking has been riddled with a wide range of attack vectors and unclear security objectives. This has led to logic locking largely remaining a theoretical concept without any tangible outcome.

To address this issue, in this book, we aim at closing the practicality gap in logic locking by devising a set of models, software tools, attacks, and schemes that enable the evaluation and application of logic locking to complex, silicon-proven HW designs within a concise and realistic attack scenario [164].

1.1 Outline

This book is organized into four parts covering eleven chapters. The structure is meant to guide the reader from basic concepts on hardware security to software implementation details for silicon-ready logic locking. By the end of the book, readers should be able to understand how logic locking operates and how it can be challenged, how to implement the right tools to deploy locking schemes, and finally, how to evaluate the security of logic locking with emerging machine learning-based approaches. The book is structured as follows.

Background First, preliminaries on electronics supply chain threats and solutions are presented in Chap. 2.

Hardware Trojans Chapter 3 introduces the anatomy of Hardware Trojans along-side existing classification systems. Moreover, a consolidated classification is introduced that considers the impact of defensive approaches. Finally, the chapter compares the effectiveness of existing Trojan-insertion countermeasures w.r.t the constructed classification.

Working Principles and Attack Scenarios The mechanics of logic locking, its impact on reverse engineering as well as common attack scenarios are discussed in Chap. 4.

Attacks and Schemes An overview and classification of deobfuscation attacks and logic locking schemes is presented in Chap. 5.

Security Metrics Chapter 6 introduces the design of one of the first generalized hardware security metrics with respect to logic locking. Furthermore, based on the introduced concepts, the security–cost trade-off problem is analyzed through a case study that evaluates the impact of a higher cost budget on the security properties of logic locking.

Software Framework The design and implementation of a software-based logic locking framework for the protection of complex multi-module HW designs is discussed in Chap. 7. The framework is designed in the form of an end-to-end locking procedure, featuring a technology-independent design representation and an extensible code base for rapid scheme prototyping. Furthermore, the constructed framework ensures the deployment of logic locking within an industry-ready setting without impacting the traditional design, verification, and fabrication steps.

Processor Integrity Protection The implementation of framework extensions in the form of two protection schemes, Inter-Lock and Control-Lock, is presented in Chap. 8. Inter-Lock embodies a cross-module, logic locking meta-scheme that scales any locking policy across multiple HW modules, thereby creating additional functional and structural dependencies between the selected components.

This cross-module policy is the first to widen the security implications of logic locking to complex hardware designs. Control-Lock implements an inter-module encryption mechanism that aims at protecting critical HW control signals against the exploitation by software-controlled hardware Trojans. The impact of both procedures is evaluated on silicon-proven RISC-V processor cores. Finally, the presented research developments are successfully transferred to industry, resulting in the first comprehensively logic-locked and commercially available processor—the "Made in Germany RISC-V" (MiG-V) core [195]. Hereby, a major milestone is achieved in the domain of logic locking.

Security Evaluation with Machine Learning The introduction of fundamental concepts in attacks and defenses in logic locking with respect to Machine Learning (ML) [170] is presented in Chap. 9. First, SnapShot is presented; an attack that utilizes artificial neural networks to directly predict correct key bits from a locked netlist. Furthermore, a neuroevolutionary procedure is developed to automatically assemble suitable neural architectures for the selected prediction problem. Furthermore, the generalized set and self-referencing attack scenario are discussed as standard attack vectors in a machine learning-based setting.

Designing Deceptive Logic Locking Based on the lessons learned from Chap. 9, the first theoretical test for uncovering structural leakage points is introduced in Chap. 10. The test embodies a procedure that can lead to the identification of fundamental security vulnerabilities that are exploitable by ML-driven attacks. The analysis results are used as a basis to construct a multiplexer-based locking policy that targets learning resilience. Through further evaluation steps, an analysis of challenges in ML-resilient locking is performed. Furthermore, a novel attack is presented, uncovering a major fallacy in existing multiplexer-based schemes. The introduced concepts, policies, and attacks offer the potential to establish the cornerstones for the design of next-generation logic locking in the era of machine learning.

Next Steps New research directions and open challenges are discussed in Chap. 11. Finally, Chap. 12 concludes the book.

Chapter 2
Background

To better understand the contents of this book, this chapter introduces the following preliminaries. An overview of the major security vulnerabilities in the electronic supply chain is presented in Sect. 2.1. Prominent Design for Trust (DfTr) solutions are detailed in Sect. 2.2. Finally, Sect. 2.3 concludes the chapter.

2.1 Electronics Supply Chain Threats

The complexity and distributed nature of the modern electronics supply chain have led to a lack of trust and assurance thereof. Consequently, a range of attack vectors has been introduced to steal, illegitimately sell, or compromise the integrity of Integrated circuits (ICs). The following subsections present the background on selected trust issues. More details can be found in [29].

2.1.1 Reverse Engineering

In the context of hardware, Reverse Engineering (RE) is defined as the process of extracting a set of specifications for a hardware design by someone other than the original design owner [136]. Hereby, RE can be deployed at different circuit abstraction levels [125]. The legitimacy of RE depends on what its result is used for. Thus, the product of RE can be utilized for either verification purposes or illegal actions, such as hardware Trojan insertion or Intellectual Property (IP) theft.

The process of reverse engineering hardware designs includes a set of manual and semi-automated steps [20, 63, 176, 187, 198]. Starting from a fabricated IC, the RE flow can be divided into netlist extraction and functionality identification. The former extracts a netlist representation of the physical chip through multiple

© The Author(s), under exclusive license to Springer Nature Switzerland AG 2023
D. Sisejkovic, R. Leupers, *Logic Locking*,
https://doi.org/10.1007/978-3-031-19123-7_2

successive steps, including the sample preparation (package removal and delayering), image acquisition, layout extraction, and netlist generation. The latter concerns the acquisition of a high-level description of the intended functionally of the design [22].

Due to its complexity, the process of reverse engineering has no clear guidelines. Only recently, first attempts to analyze the required cognitive and technical skills to perform RE have been analyzed [25]. Nevertheless, it still remains a challenge to fully automate the process as well as quantify the complexity of RE for a specific design.

2.1.2   Hardware Trojans

Hardware Trojans (HTs) are defined as malicious and intentional circuit modifications that can result in undesired circuit behavior after deployment [30, 201]. The malicious behavior can be manifested in the form of information leakage, performance degradation, increased power dissipation, Denial of Service (DoS) attacks, and others.

The anatomy of hardware Trojans consists of a trigger and a payload [28, 39]. The trigger activates the Trojan based on a specific activation event, such as the occurrence of specific data values or circuit states, external signals, number of cycles, and others. The malicious behavior of the Trojan is manifested in the form of the payload. The malicious circuitry can be inserted into the hardware at different design levels, depending on which entities in the supply chain are considered trustworthy. Thus, in principle, HTs can be introduced into a design during specification, design, fabrication, testing, or assembly and packaging, thereby being initiated by untrusted personnel or Electronic Design Automation (EDA) tools. This broad attack landscape has led to the introduction of many hardware Trojan taxonomies and example implementations [51, 90, 143, 148, 186, 208]. Furthermore, a recent study even demonstrated how hardware Trojans can effortlessly and automatically be implanted in finalized layouts [120].

The diverse design possibilities, insertion locations, and stealthy implementation nature make HT detection a challenging task. Moreover, similar to regular faults, the later Trojans are detected in the design and production flow, the costlier and more difficult it becomes for the IP owner to act. This problem is further exacerbated by the fact that HTs are often assumed to be deployed by untrusted, external foundries— beyond the control of the legitimate IP owner.

Another important aspect lies within the resources required to design, implement, and inject hardware Trojans [72, 154]. In general, *design-independent* HTs can be implemented and inserted with very little knowledge about the design's functionality or structure. These Trojans, however, are likely to be detected, exhibit uncontrollable trigger mechanisms, and result in random payloads, thus mimicking a random fault. In contrast, *design-dependent* Trojans can be constructed to allow for a controllable activation, stealthy implementation, and dedicated payload, thus leading to high-impact attack scenarios. Consequently, design-dependent HTs require the attacker to reverse engineer and understand the design's functionality and structure to successfully construct and insert the malicious modifications [218, 223]. More details on hardware Trojans are presented in Chap. 3.

2.1.3 IP Piracy and Overuse

IP piracy refers to the illegal or unlicensed use of intellectual property [218]. The act of piracy can occur at different points in the electronic supply chain. For example, a dishonest System-on-Chip (SoC) designer might make illegitimate copies of a legally purchased Third-Party Intellectual Property (3PIP) to sell to other SoC designers. Moreover, the SoC designer can perform certain modifications and sell the altered design as new IP. Piracy can occur at later design and fabrication stages as well. For example, if an IP design is outsourced to external design houses or foundries, malicious actors can sell illegal copies of the IP (in netlist or layout format) [29].

Another threat is known as IP overuse, where an entity uses the IP in more instances than contractually agreed upon [15]. In general, the IP legally belongs to the IP author. SoC integrators often rely on purchasing IPs from legitimate IP owners to design a chip. Thus, IP owners can license the integration of their IPs to the SoC designer for a specific number of chips. However, a rogue SoC designer might overuse the IP by issuing the production of more chips beyond the agreed licensing scheme, thus generating losses for the IP owner. Note that the semiconductor IP market was valued at $5.33 billion in 2019 and is projected to reach $7.44 billion by 2027 [191]. Thus, genuine IP owners have a strong economic incentive to protect their assets.

2.1.4 IC Overbuilding and Counterfeiting

Untrusted foundries and assembly facilities can produce more chips than legally contracted, resulting in IC overbuilding. This trust issue arises out of the inability of the IP owner and the design house to monitor the fabrication and assembly process. The malicious entities can overbuild chips at a high profit as no design development cost is required. Moreover, overbuilding can be performed at virtually no cost by reporting a lower yield [29, 139].

In this context, a serious threat is posed by IC counterfeiting. Counterfeit ICs are replicas of the genuine integrated circuit that functionally and visually appear identical to the genuine ICs. These include remarked, out-of-spec, or recycled ICs, often extracted from discarded electronic devices [218]. Besides incurring revenue losses to legitimate IP owners, overbuild and counterfeit components are a major source of serious reliability and security issues. With minimal or no testing and verification, these ICs end up having an impact on personal computers, telecommunications, automotive electronics, medical devices, and military systems [64, 65]. For example, the Pentagon—the highest military headquarters of the United States—has reported that about 15% of all its purchased electronic parts are counterfeit. Moreover, it is estimated that the counterfeiting of electronic parts incurs an annual loss of more than $7.5 billion for legitimate semiconductor manufacturers [144].

2.2 Design-for-Trust Solutions

Over the last decades, a range of DfTr solutions has been proposed to enforce security and trust throughout the IC supply chain. In the following, we present a subset of these solutions that focus on *hardware integrity protection*.

2.2.1 Layout Camouflaging

Layout Camouflaging (LC) aims at preventing reverse engineering by replacing selected cells with their camouflaged counterparts [130]. The camouflaged cells have a structurally similar construction across different gate types. Thus, from the RE perspective, the attacker has to invest further efforts to uncover the correct functionality. LC can be performed through obfuscating the interconnects [119], using dummy contacts [130], threshold voltage-dependent cells [61], or filler cells [42], combining doping modifications and dummy contacts [149], and leveraging AND-tree camouflaging [102], among others. A significant drawback of the LC methodology is the reliance on a trusted foundry to implement the camouflaged cells. Thus, in its current state, LC can only protect against post-fabrication RE. Nevertheless, recent work has investigated the application of polymorphic spin devices for the implementation of dynamic camouflaging [133]. This novel approach to camouflaging attempts to extend the protection mechanism to cover untrusted foundries as well.

2.2.2 Split Manufacturing

Split manufacturing is a methodology that attempts to protect against an untrusted foundry by splitting the layout into two parts: the Front End of Line (FEOL) and Back End of Line (BEOL) [83]. Each part is fabricated in separate, independent foundries. FEOL includes the transistors and lower metal layers ($\leq M3^{1}$), and BEOL includes the remaining higher metal layers ($\geq M4$). In this setting, the low-cost BEOL layers can be fabricated in a trusted in-house foundry and only the more expensive FEOL layers need to be fabricated in an untrusted, high-end foundry. Thus, only the

---

[1] Mx denotes the metal layer x. Note that there is no exact splitting rule.

transistors and a limited set of connections are exposed to the untrusted party [212]. Split manufacturing offers a potential way to simultaneously reduce the cost of in-house production and lower the security risk related to outsourcing the IC fabrication [131].

2.2.3 Metering

Hardware metering refers to a set of protocols and methodologies that allow the IP owner to track the manufactured ICs after fabrication, thus enabling the prevention and detection of overbuild and counterfeit ICs [6, 95]. Hardware metering can be categorized into *passive* and *active* approaches. Passive metering can uniquely identify ICs through indented and digitally stored serial numbers, unclonable identifiers, and others [94, 109]. Therefore, suspect ICs can be taken from the market and checked for a valid registration. A major limitation of passive metering is that it can only detect pirated ICs without being able to actively counteract piracy. Active metering, on the other hand, locks each produced IC until it is activated by the IP owner [43, 66]. In this sense, active metering can be implemented by using different combinational or sequential logic locking mechanisms. However, we categorize active metering as a protocol, not a specific technique. Thus, it is not included in further comparisons.

2.2.4 Functional Filler Cells

The unused spaces in the chip layout are typically filled with standard (non-functional) filler cells to increase the uniformity of the circuit density. However, as these serve no functional purpose, filler cells provide a suitable location to insert hardware Trojans; especially, as removing these cells has a small impact on the chip's electrical parameters [29]. This threat can be actively mitigated by utilizing *functional* filler cells. One such approach is known as Built-In Self-Authentication (BISA) [21, 202, 203]. The principal idea of BISA is to connect all filler cells to form a testing circuitry. After production, the BISA circuit can be tested for any potential changes in the connected cells. Thus, BISA can uncover potential malicious modifications, such as layout-level hardware Trojans, that change the initial design of the functional filler cells.

2.3 Synopsis

This chapter introduced the background on the predominant security and trust vulnerabilities and solutions within the electronics supply chain.

Chapter 3
Hardware Trojans

Classifying hardware Trojans is certainly a challenging and, sometimes, never-ending task. The limitless design variability makes it increasingly difficult to capture all necessary features into a single classification. To elaborate on these challenges, in the following, we describe the basic anatomy of Hardware Trojans (HTs) and introduce background on existing classification proposals in Sects. 3.1 and 3.2, respectively. A new classification system, which takes defensive approaches into account, is discussed in Sect. 3.3. A comparison of the capabilities of existing defensive approaches w.r.t. the newly introduced Trojan classification is presented in Sect. 3.4. Finally, Sect. 3.5 concludes this chapter.

3.1 The Anatomy of Hardware Trojans

Regardless of the exact implementation details, the anatomy of a hardware Trojan can conceptually be separated into the *trigger* and *payload* circuitry, as shown in Fig. 3.1. The trigger activates the HT based on a specific input event. This event might be driven by internal circuit signals or an external source. Typically, it is assumed that a trigger might be sensitive to a selected data value, circuit state, a number of processing cycles, physical event (e.g., temperature change), and others. The intended malicious act is manifested in the form of the payload that can, in principle, lead to any desired attack vector—from a slight disturbance to a disastrous fallout [90].

© The Author(s), under exclusive license to Springer Nature Switzerland AG 2023
D. Sisejkovic, R. Leupers, *Logic Locking*,
https://doi.org/10.1007/978-3-031-19123-7_3

Fig. 3.1 The anatomy of hardware Trojans

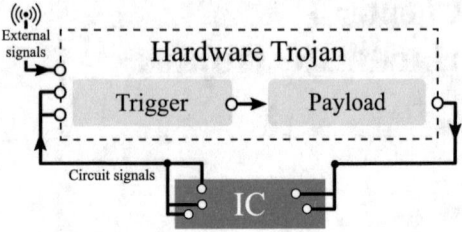

3.2 Classifications

Throughout the past decades, a few attempts have been made to establish a common classification system for hardware Trojans. These classifications or taxonomies aim at compartmentalizing HTs based on a set of classification rules. We discuss two popular systems in the following.

3.2.1 Activation and HT Effect

A high-level classification with respect to the HT activation mechanism and effect was proposed in [28]. This taxonomy classifies all HTs based on the *trigger* and *payload* design, as shown in Fig. 3.2a. The first trigger class includes all digital designs that are sensitive to a specific Boolean function. These can further be separated into combinational and sequential implementations. The second trigger class describes analog designs that are sensitive to analog conditions such as delay, temperature, device aging, and others. The payload design is categorized into three classes: digital, analog, and others. These classes describe a plethora of attack vectors, including functional failures, information leakage, die overheating, and enabling side channels.

3.2.2 Comprehensive Classification

A comprehensive hardware Trojan taxonomy has been introduced and successively refined in [90, 148, 186]. The taxonomy is visualized in Fig. 3.2b. This model is based on multiple categories, as discussed in the following. The *insertion phase* describes Trojans based on the Integrated Circuit (IC) supply chain stage in which the Trojans have been introduced, ranging from specification to packaging. The *abstraction level* captures the design level in which a Trojan has been injected, starting from the system level down to the physical implementation. The *activation mechanism* classifies Trojans based on their activation nature, including always-on functions and condition-based triggering. The Trojan *effect* categorizes HTs based

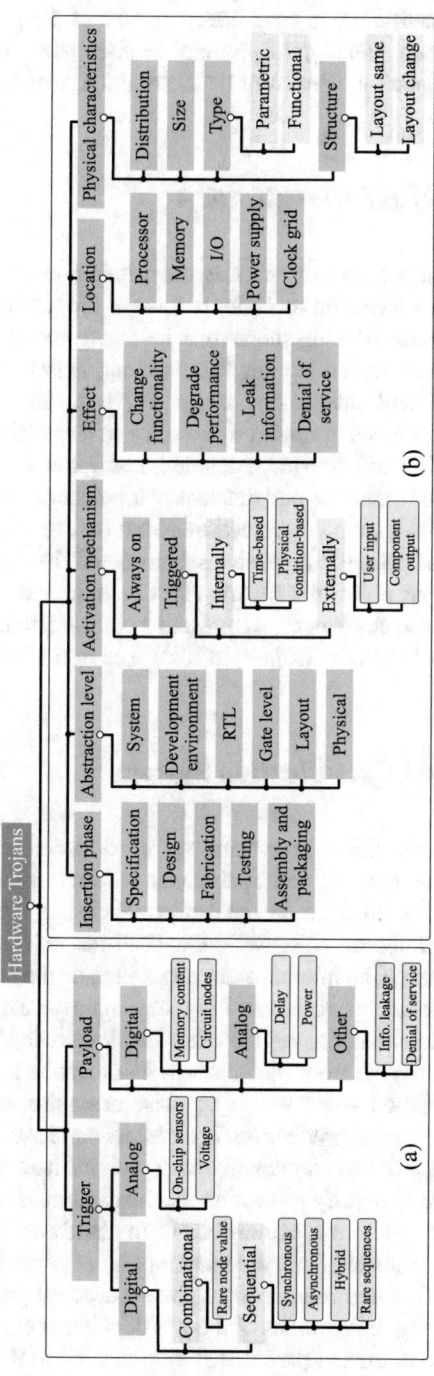

Fig. 3.2 Hardware Trojan classification based on (**a**) activation and HT effect and (**b**) a plethora of HT design features

on their malicious impact on the underlying design. The *location* describes Trojans based on which component the HTs have infected, such as the processor, memory, or clock grid, among others. Finally, the *physical characteristics* classify HTs based on the physical implementation in terms of HT distribution, size, type, and structure.

3.2.3   Challenges of HT Classification

The discussed classification systems aim at providing a detailed description of HT implementations. However, these systems have a few critical disadvantages. First, at their core, HTs are nothing more than hardware modifications, and thus a practically limitless sea of Trojan implementations exists. Consequently, the presented classifications will have to constantly evolve in time. The second, even more critical, issue is the missing notion of security in terms of countermeasures. Even though the classifications provide a detailed description of potential Trojans, these do not express to which extent defensive approaches can protect against specific hardware Trojans. Having a protection-aware HT classification is a critical tool for the evaluation of potential defensive schemes and the concise modeling of attack scenarios—a major point of confusion in scientific and industrial security proposals. To address these challenges, we present a consolidated, protection-aware classification system for hardware Trojans, as discussed in the following section.

3.3   A Consolidated Classification System

Before introducing the classification system, we need to define which assets are being protected. In principle, we can distinguish between *internal* and *external* assets. Internal assets are intrinsic to the hardware design itself, including the design's functionality, structural composition, and all the information that can be learned thereof. For example, an internal asset might include the fact that a hardware component embeds a processor design and cryptographic accelerators or hosts an Ethernet controller. External assets, on the other hand, include all the information that is not physically derivable from the hardware but must be kept secret. This is, for example, the information about which use case or environment the hardware component is deployed in after production. Both assets can have an impact on how a hardware Trojan is constructed and how it can exploit the hardware design.

Currently, all hardware integrity protection mechanisms are focusing on protecting internal assets, since these are exploitable in the hardware itself. In contrast, external assets must be protected from the human point of view. For example, if an Intellectual Property (IP) owner intends to produce a secured processor design for military aviation, evidently the specifics of the IP's deployment environment should not be communicated to untrusted parties that are involved in the IC supply chain.

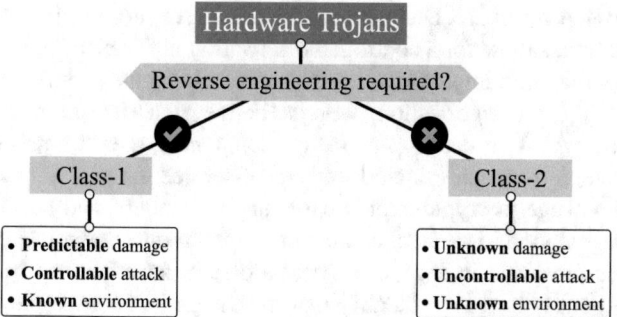

Fig. 3.3 Hardware Trojan classification based on the reverse engineering requirements

In the rest of the discussion, we assume that external assets are inherently protected, thus offering no ground for hardware Trojan design and insertion.

To design a Trojan classification system from *a defender's point of view*, we have to first identify the key differentiator that has an impact on the efficacy of all defensive approaches with respect to HT prevention. This role is played by the amount of required Reverse Engineering (RE) effort to perform Trojan design and insertion. Hereby, we define RE as the process of understanding the functionality, structural composition, and potential deployment environment of the design under attack only based on *the internal assets*. The RE process is nowadays compiled of a set of complex manual and semi-automated steps, typically consisting of two stages: netlist extraction and functionality identification [125, 198]. Netlist extraction acquires a netlist representation of the physical chip through a series of stages, including package removal and delayering, chip imaging, layout extraction, and netlist generation. Functionality identification attempts to extract a high-level description of the initial functionality of the analyzed design. Henceforth, the design under attack is referred to as the *target design*.

Regardless of the HT type, a malicious entity has only two options to insert a Trojan. The first option is to insert Trojans for which *at least some RE is necessary*. In contrast, the second option is to opt for Trojans for which any form of RE *is not required*. The amount of RE-based effort involved in Trojan design and insertion defines the potential efficacy of a defensive approach. Using this concept, we can construct a radically simplified classification system that separates all HTs into two possible classes, as described in the following and visualized in Fig. 3.3.

3.3.1   Class-1 Hardware Trojans

The first class defines Class-1 hardware Trojans (C1HTs) that are *RE-dependent*. This has two consequences. First, a certain effort must be invested by the malicious entity to understand the target's specifications to be able to draft and insert a design- and environment-dependent HT. Second, because the C1HT is designed specifically for a given target, it can allow for a controllable activation mechanism and the execution of a very specific, high-impact attack—thus, maximizing the incurred damage.

To simplify the categorization, we can define all C1HTs as malicious modifications for which *at least one design component*, i.e., trigger or payload, is RE-dependent. Examples include denial-of-service HTs in a missile or car, information leakage in cryptographic cores in card readers, and performance and reliability degradation in medical equipment—among many others. A simple gate-level C1HT is presented in Fig. 3.4a. After a certain RE effort has been invested, the untrusted entity is able to identify that the design is, for example, a processor core. Furthermore, within the core, the exact position of the Arithmetic Logic Unit (ALU)-enable signal is located. This signal controls if the ALU is enabled or disabled for operation execution. In this case, the ALU-enable signal can be manipulated by the sample HT. For simplicity, both the trigger and the payload are implemented using a single gate. The trigger reacts to the input event $i_1 = 1$ and $i_2 = 1$. If activated, the payload inverts the value of the signal x. Consequently, the HT leads to a reliability-degradation attack in which the processor is unable to correctly execute all instructions in every cycle. Put in the right context, such as in a nuclear power plant, this HT can have catastrophic consequences.

Note that the deployment environment can play a crucial role in Trojan design as well. For example, knowing whether the final chip is deployed in a mobile phone, nuclear power plant, or a closed military compound can have a profound impact on the Trojan's trigger or payload construction as well as the profitability of the attack. Hereby, we assume that environmental information is gathered based on internal assets, thus being conditioned by the RE process.

Since this Trojan class requires a certain amount of invested RE effort from the untrusted entity, any defense that *increases the complexity of RE* can be effective against C1HTs. For example, logic locking, a hardware obfuscation technique, induces functional and structural design changes that lead to a more complex path toward understanding the design [218]. Therefore, logic locking increases the effort and cost for an attacker to design and insert a C1HT. Note, however, that the concrete effectiveness of a specific scheme relies on many different factors. Nevertheless, any change that increases the complexity of the initial target design indisputably increases the required RE effort.

3.3.1.1 Example of a Class-1 HT in Processors

To better understand the described classification and anatomy of hardware Trojans, let us consider one more example of a C1HT in Fig. 3.5. Here, a simple gate-level HT is constructed to perform a Denial of Service (DoS) attack within a processor that is responsible for tracking the geographical location of a vehicle, such as an airplane or a container ship. The HT is triggered once a selected geographical location is reached. For simplicity, only the relevant components of the microarchitecture are visualized. The HT is designed as follows. The HT trigger continuously tracks the

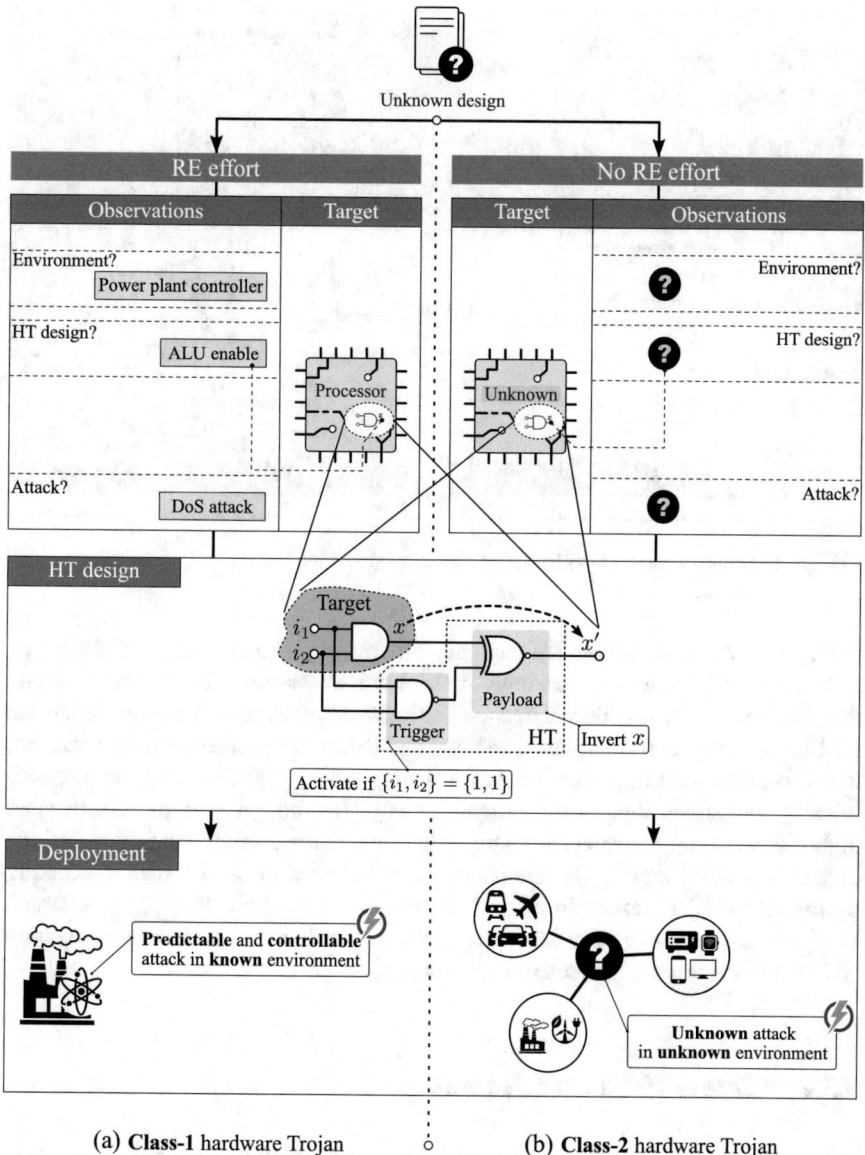

(a) **Class-1** hardware Trojan   (b) **Class-2** hardware Trojan

Fig. 3.4 Example of a gate-level (**a**) class-1 and (**b**) class-2 hardware Trojan

two data output ports of the register file within the processor. If the two data values match a predefined latitude and longitude, the trigger's comparison unit outputs the value 1. This value is further routed through a single-bit register to ensure that the value remains 1 in all further cycles. Finally, this value is forwarded to

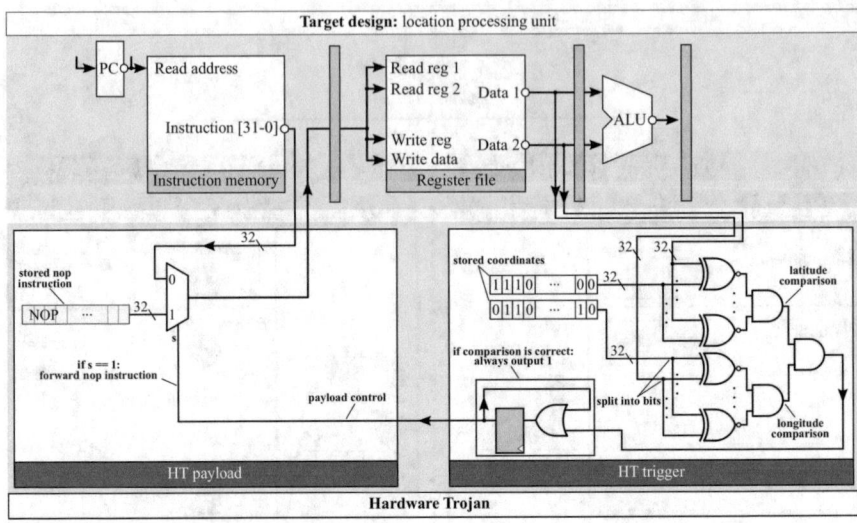

Fig. 3.5 Example of class-1 DoS hardware Trojan on gate level

the controlling input of the HT payload. The payload stores a copy of the nop[1] instruction and implements a single multiplexer to choose between the nop and the next fetched instruction based on the controlling input. Therefore, once the HT trigger consistently outputs 1, the payload forces the processor to only execute the nop instruction in all the following cycles—effectively disabling the location-tracking unit. Even though this simple HT is of low complexity, its effectiveness highly depends on the attacker's ability to identify the correct instruction set, the processor's components, the intended use case, as well as the data processing details, such as the exact format of the coordinates within the target software. Hence, successfully reverse engineering the design is a predicate to the insertion and deployment of the presented design-dependent C1HT.

3.3.2   Class-2 Hardware Trojans

The second class defines Class-2 hardware Trojans (C2HTs) that are *RE-independent*. The design and insertion of C2HTs are, therefore, completely independent of the target's specifications and deployment environment. Thus, compared to the first class, the insertion of C2HTs is a low-cost procedure, because no RE effort is required. Moreover, a C2HT-enabled attack is, on average, most likely of a low impact as the inserted Trojan is injected into the target without knowledge about

---

[1] nop is an instruction that performs no operation.

its functionality, structure, or use case. Therefore, the likelihood that a C2HT remains undetected and that it exploits circuit components that result in a disastrous, targeted, high-impact attack is low.

In terms of categorization, we can characterize all C2HTs as malicious modifications for which both *the trigger and the payload* are RE-independent. Some examples include random fault injections, random net widening or narrowing to degrade performance, insertion of ring oscillators to increase power usage, manipulating the physical characteristics of transistors, and many others.

As the insertion of C2HTs is not predicated on the success of the RE process, any defensive scheme that impacts the RE complexity will most likely not be able to protect against this Trojan class. An example of a gate-level C2HT is shown in Fig. 3.4b. Since no RE effort is invested, the malicious entity has not retrieved any specifications of the target design. Thus, a randomly selected wire is manipulated with the presented gate-level HT. Even though the HT implementation is equivalent to the class-1 case, the certainty of a meaningful attack outcome is heavily predicated by the choice of the design-specific internals to be manipulated. Consequently, the presented C2HT sporadically injects faults into the target IC. Evidently, even if the IC is highly protected, e.g., through logic locking, and the untrusted entity is completely unaware of the target's functionality or use case, the Trojan insertion *is still possible*.

3.3.3 Classification Features

The advantages of the proposed classification system are manifold. First, the basis on which the hardware Trojans are classified fully transcends any implementation details, such as the trigger and payload type, insertion phase, abstraction level, location, physical characteristics, and others. Therefore, the system does not suffer from the necessity to predict all possible HT implementations. Second, the simplicity of the model allows for a precise definition of an attack scenario. For example, the development of a logic locking scheme is naturally motivated by the protection against C1HTs, regardless of the true effectiveness against potential deobfuscation attacks. In contrast, the existing classification proposals are not able to express the protection coverage of such defensive approaches, leading to vague and incomplete attack formulations. Finally, even though it is not in the focus of our discussion, the two-class system can also clearly suggest which Trojan types are more resilient against detection. Since C1HTs are specifically drafted for a target design, these Trojans are likely to be more stealthy and avoid detection mechanisms. C2HTs, on the other hand, are more likely to be filtered out during testing, activation, or IC power-up, as their design might be bound to parts of the target which lead to clearly identifiable changes—a consequence that is not evident without sufficient RE results.

3.4   Preventing Hardware Trojans

In the past decades, many Design for Trust (DfTr) techniques have been proposed to enforce trust throughout different stages of the IC supply chain, as discussed in Chap. 2. Some of these proposals can potentially be utilized to protect against hardware Trojans. Thus, in the following, we analyze the effectiveness of standard DfTr methodologies in thwarting hardware Trojans based on the consolidated classification system, following the summary in Table 3.1. Note that we can only indicate the *potential* of a particular DfTr method to protect against HTs; its exact efficacy, however, is subject to further evaluations in the particular use case.

3.4.1   Layout Camouflaging

This DfTr technique aims at increasing the complexity of RE through the replacement of circuit cells with their camouflaged counterparts [130]. The camouflaged cells have a structurally similar layout for multiple gate types, thus forcing the attacker to invest more effort to correctly identify the functionality of a design. As layout camouflaging relies on the collaboration with a trusted foundry, it can only protect against entities after fabrication. Furthermore, since layout camouflaging increases the RE complexity, it can protect against C1HTs. However, it has no impact on C2HTs as their insertion is not hindered in any aspect.

3.4.2   Split Manufacturing

Split manufacturing attempts to protect against an untrusted foundry by separating the layout into the Front End of Line (FEOL) and Back End of Line (BEOL) [212]. The two parts are fabricated in separate foundries. FEOL includes the lower metal layers and the transistors, whereas BEOL includes the remaining higher metal layers.

Table 3.1 Design-for-trust techniques for the protection against hardware Trojans. Each table entry specifies the impact on the proposed HT classes as follows: C1/C2

DfTr method	External design house	Foundry	Assembly facility
Layout camouflaging	✗/✗	✗/✗	✓/✗
Split manufacturing	✗/✗	✓/❖	✗/✗
Functional filler cells	✗/✗	✓/❖	✓/❖
Logic locking	✓/❖	✓/❖	✓/❖

✗ marks that a technique **cannot protect** against attacks by the untrusted entity
✓ marks that a technique **can protect** (to some extent) against the untrusted entity
❖ marks that a technique **might be able to protect** against the untrusted entity **for at least some Trojans**

The low-cost BEOL is typically fabricated in a low-end, in-house foundry. Thus, only the expensive FEOL layers are fabricated in a high-end, untrusted fabrication facility. Since the untrusted foundry only acquires part of the design, split manufacturing can potentially increase the required RE effort, thus having an impact on the insertion of C1HTs by the untrusted foundry. Moreover, since the untrusted foundry would have to insert Trojans which result in the correct functional behavior of the produced IC after the FEOL and BEOL layers are integrated, more complex C2HTs might be easily uncovered. However, transistor-level C2HT manipulations are still a viable option. Therefore, the role of split manufacturing must further be evaluated for C2HTs. As the initial circuit layout remains functionally and structurally unchanged, the complexity of RE is not increased after fabrication. Thus, no protective notion is provided against assembly-level Trojan insertion.

3.4.3 Functional Filler Cells

Standard filler cells are utilized to fill the unused spaces in the chip layout to increase the uniformity of the circuit density. To hinder the insertion of Trojans in these empty layout spaces, functional filler cells can be deployed [202]. These cells can be connected to form a testing circuitry. By testing the functional filler cells after production, a manipulation of the cells can be detected. Moreover, the correct identification of functional filler cells requires an additional effort from the attacker. Consequently, functional filler cells can help in protecting against foundry-level C1HTs. The same is true for assembly-level Trojans, as the assembly facility is also forced to invest more RE effort. Similarly to split manufacturing, functional filler cells might help in protecting against C2HTs if these tend to manipulate the empty spaces in a layout.

3.4.4 Logic Locking

Logic locking is a DfTr technique to protect the integrity of hardware designs at different stages of the IC supply chain [218]. Logic locking alters hardware designs by inducing functional changes that are bound to an activation key, which is only known to the IP owner. Since the locked designs are functionally and structurally changed through the insertion of additional logic and the adaptation of existing functionality, the complexity of the RE process is increased. Moreover, since logic locking is typically deployed at gate level, the RE effort also increases for an untrusted external design house. Consequently, the protective impact of logic locking ranges from the external design house to the assembly facility. Logic locking might have an impact on some C2HTs as well. For example, if a C2HTs is by chance inconveniently placed in the activation mechanism of logic locking, it might lead to the inability of activating the IC properly, thus resulting in a disregarded chip, even before it can reach its intended deployment environment.

Other DfTr techniques, such as runtime monitoring and HT-tolerant computing, are beyond the scope of this article, as these do not prevent Trojan insertion [29].

3.4.5 Lessons Learned

The presented comparison of DfTr techniques opens new perspectives in the prevention of hardware Trojans. In the context of C1HTs, a sufficiently complex RE procedure might disable the ability of an attacker to insert a hardware Trojan, especially since the attack window is typically limited in time. A fundamentally important question is, however, still left unanswered: how much RE is required to gather sufficient information for a successful C1HT design and insertion? The capability of specific DfTr techniques to protect against C1HTs has long been only a theoretical assumption. Their true RE-based security measure has yet to be explored.

Moreover, in terms of C2HTs, we have established that their insertion is not predicated on RE. Therefore, it is extremely difficult to efficiently protect against this Trojan type. However, based on the previous DfTr comparison, we can extract a few observations that point to cases when existing defensive approaches could, potentially, be effective. For example, C2HTs might be more easily discovered if their insertion is done before the complete physical structure of the chip is known to the attacker. This is the case in split manufacturing where an attacker might place a C2HT that leads to a faulty IC right after integration, simply because the Trojan already corrupted the integration itself. A similar observation can be made in terms of logic locking. A Trojan that unintentionally disables the activation process can lead to a chip being disregarded even before its deployment.

3.5 Synopsis

The limitless implementation possibilities of hardware Trojans make their classification an inherently difficult problem. Existing proposals often stumble over the variable Trojan nature leading to the inability to express the impact of defensive approaches in preventing different Trojan types. To address this issue, we have presented a consolidated classification system that categorizes Trojans from a defender's point of view. By dividing hardware Trojans based on their relation to reverse engineering, the proposed classification perspective radically simplifies the Trojan categorization, thereby enabling a clear notion of the defensive effectiveness of design-for-trust techniques as well as the concise modeling of attack scenarios. Furthermore, the proposed classification has been analyzed in the context of contemporary defensive techniques. Finally, gathered observations have been discussed, laying out new perspectives in hardware Trojan prevention.

Part II
The Mechanics of Logic Locking

Chapter 4
Working Principle and Attack Scenarios

As elaborated in the previous chapter, compared to other Design for Trust (DfTr) solutions, logic locking has the potential to protect against untrusted entities throughout the Integrated Circuit (IC) supply chain [15, 34, 221, 223].[1] The core mechanism of logic locking relies on performing design manipulations by binding the correct functionality of a hardware design to a secret activation key that is only known to the legitimate IP owner. If the correct key is provided, the design performs as originally intended. Otherwise, an incorrect key ensures that the circuit generates faulty output values for at least some input patterns. Alongside the functional implications, logic locking also induces structural hardware changes. Thus, the original functionality and structure (topology) of the hardware design remain concealed while passing through the hands of external, untrusted parties. As logic locking can protect against a range of untrusted parties within the IC supply chain, this technology is placed at the core of this book for the protection of the integrity of complex hardware designs. The core mechanics of logic locking are discussed within this chapter alongside the attack scenario assumed in this book. The introduced basics should prepare the reader for the next chapters.

4.1 Classification

Logic locking can be broadly classified into two orthogonal classes: *combinational* and *sequential* [154]. Combinational logic locking performs key-dependent manipulations in the combinational path of the hardware design [223]. Sequential locking, on the other hand, concerns with the obfuscation of the state space of the circuit [38, 50, 113]. Furthermore, logic locking can be categorized based on its

---

[1] Note that logic locking can theoretically also protect against Intellectual Property (IP) overuse and IC overproduction [54].

© The Author(s), under exclusive license to Springer Nature Switzerland AG 2023
D. Sisejkovic, R. Leupers, *Logic Locking*,
https://doi.org/10.1007/978-3-031-19123-7_4

operational level, including algorithm-level or Register-Transfer Level (RTL) logic
locking [79, 87, 123], and gate-level logic locking [54].

Thus far, the mainstream of research has been focusing on combinational
gate-level logic locking as it enables a wide range of cost-effective scheme
implementations and a simple means of deployment. This trend is followed in this
book as well. Thus, henceforth, the term *logic locking* refers to *combinational gate-
level* locking schemes.

4.2 Locking Example and Notation

To demonstrate the working principles of logic locking and showcase the basic
notations used in later chapters, let us consider the simple example in Fig. 4.1a.
The example presents a gate-level netlist using the IEEE standard graphic symbols
for logic functions [77]. An overview of all gate types is available in Appendix A.1.
Four distinct components can be identified in the netlist: primary inputs, primary
outputs, logic gates, and internal signals. For a selection index j, these components
are defined as follows:

- Primary inputs are 1-bit signals that can be externally controlled, i.e., a selected
  input vector can be provided. A primary input is defined in the form of i_j.
- Primary outputs are 1-bit signals that can be externally observed. A primary
  output is defined as o_j.
- Logic gates are represented using the standard notation and identified as g_j.
- Internal signals (gate outputs and inputs) are denoted as s_j. The value of internal
  signals is not directly observable.

Note that sequential circuits must provide scan chain access to control or observe
primary signals within the circuit.

An example application of logic locking based on XOR and XNOR gates is
shown in Fig. 4.1b. Here, the locking scheme has inserted additional XOR and
XNOR gates. These gates are known as *key gates*, as they are controlled by a set
of key inputs. Thus, logic locking extends the gate-level netlist by adding a set

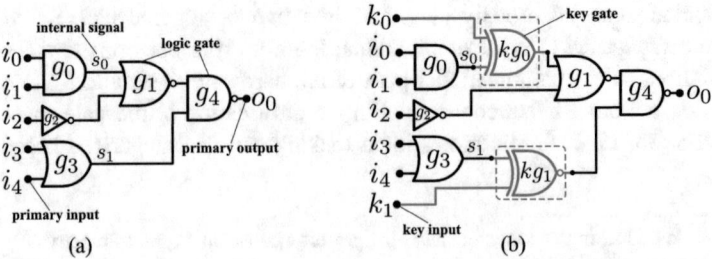

(a)   (b)

Fig. 4.1 Example of XOR/XNOR-based logic lock: (**a**) original netlist and (**b**) locked netlist

of key inputs. Each 1-bit key input is denoted as k_j.[2] Key inputs have the same properties as primary inputs. In this example, k_0 and k_1 control the key gates kg_0 and kg_1, respectively. Only if the correct key input $K = \{k_0, k_1\}$ is provided, the locked netlist performs as originally intended. For example, for $K = \{0, 1\}$, the values of the internal signals s_0 and s_1 are buffered through the respective key gates, thus preserving the functionality of the circuit. In addition to the functional dependency, the structural alteration induced by logic locking is given through the inserted key gates. Based on this working principle, a plethora of locking schemes has been introduced in the last decade. More details are provided in Chap. 5.

4.3   Logic Locking and Hardware Trojans

As discussed in Chap. 3, it is assumed that reverse engineering must be performed to implement and insert high-impact, design-dependent hardware Trojans. Due to the impact of logic locking on the design's functionality and structure, the complexity of the reverse engineering process is significantly increased. Thus, an untrusted entity has to first unlock the design by finding the correct activation key to perform the Trojan insertion.

4.4   Logic Locking in the IC Supply Chain

The role of logic locking for hardware integrity protection in the IC supply chain is demonstrated in Fig. 4.2. Note that some details, such as verification steps, are not shown. The IP owner represents the trusted entity that is introducing a legitimate IC on the semiconductor market. In the first stage, an RTL design is compiled based on the system specifications (Fig. 4.2a). Hereby, a set of third-party IPs can be integrated into the design. Furthermore, the RTL design is logic synthesized and locked, resulting in a secret activation key and a locked netlist. Both these steps are performed in a trusted in-house setting. Note that the locked gate-level netlist is typically *resynthesized* after locking to further integrate any locking-induced changes. Hence, we can differentiate the *pre-resynthesis* and *post-resynthesis* netlist. At this point, the locked netlist can proceed in the untrusted stages of the supply chain. This typically includes an external design house for the physical (layout) design (Fig. 4.2b), the foundry (Fig. 4.2c), and the assembly facility (Fig. 4.2d). Note that the assembly facility might also play an important role in hardware Trojan insertion, even though it operates only after the IC is fabricated [90]. For example, the facility might be able to insert an external malicious modification that is specific to the design. Finally, after packaging, the IC is returned back to the trusted IP

---

[2] Sometimes we refer to k_j in the form of $K[j]$ when considering the key array K.

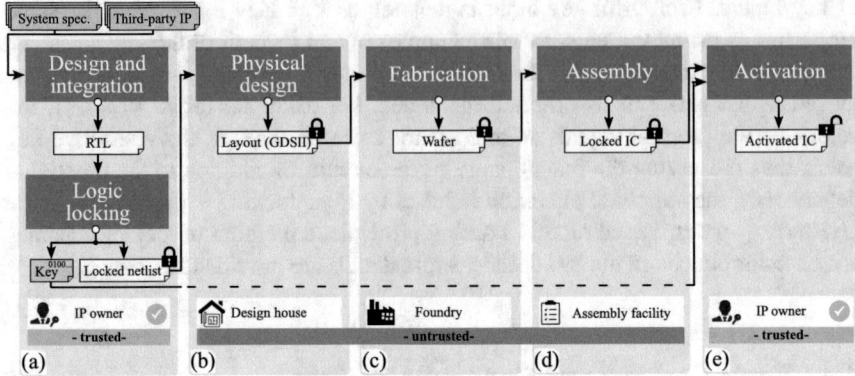

Fig. 4.2 Logic locking in the IC supply chain

owner for activation (Fig. 4.2e). Hereby, the correct key is configured through a non-volatile memory, e.g., e-fuse, flash, or EEPROM [126]. Since the activation key is only known to the IP owner, logic locking protects the design integrity by bridging the untrusted supply chain stages.

Note that the scope of the untrusted stages depends on the assumed attack model. For example, the physical chip design can be considered trustworthy if done in an in-house environment. Moreover, end users can also be seen as potentially malicious and untrusted entities. However, these are out of the scope of this book.

4.5   The Concept of Secrecy

Logic locking has two fundamental impacts on a design. First, it can render a fabricated IC inoperable due to the functional implications of the correct activation key on the correctness of the IC's operation. However, achieving inoperability can be done with much simpler methods without the involvement of logic locking [156]. The second objective of logic locking is to conceal the design's functionality. This property is also known as *functional secrecy* [156, 229]. In the context of this book, logic locking is viewed as a means to protect against hardware Trojans that rely on the correct identification of the functional secrecy of the underlying HW design. Nevertheless, the security of logic locking is mostly measured through resilience against various key recovery attacks. Thus, the key accuracy acts as a proxy to functional secrecy.

4.6 Terminology

Logic locking has both a functional and a structural impact on the design. Thus, since its inception, it has also been referred to as logic encryption or obfuscation. Through the last decade, a clear consensus on the naming has not been reached. However, lately, the scientific community has favored the term "logic locking."

4.7 Attacks on Logic Locking

As elaborated in Chap. 5, the security of logic locking has been thoroughly challenged in the past decade through different key recovery (deobfuscation) attacks. These attacks try to reconstruct the correct activation key based on the following assumptions [126, 223]:

- The attacker has access to the locked design, typically in the form of a gate-level netlist.
- The location of the key inputs (pins) is known.
- The deployed locking scheme is known.
- The attacker has access to an activated IC to use as *oracle* for retrieving golden input/output (I/O) patterns.

The existence of the last assumption categorizes all key-retrieval attacks into Oracle-Guided (OG) and Oracle-Less (OL) attacks. OG attacks assume the availability of an activated IC. This is typically the case in a high-volume production setting in which the attacker is able to acquire an activated IC instance of the design on the semiconductor market. Moreover, in certain scenarios, an attack requires the availability of only a few golden I/O patterns, e.g., in the form of test vectors. Nevertheless, this also falls into the oracle-guided attack model. On the contrary, OL attacks assume that an activated IC instance is not available. Thus, the attacker has only access to the locked design. This is often the case in a low-volume production of devices with unique and highly confidential hardware requirements, such as in security-critical and defense applications [49].

Moreover, recent findings suggest that OL attacks might have a more important role compared to the OG model in realistic attack scenarios. The main cause of this discrepancy lies within the problem of protecting the locked circuit after production is done, i.e., in deployment. These findings are as follows:

- A large set of OG attacks rely on having access to the internal combinational paths and circuit states of a design. Otherwise, the attacks become inefficient and impractical. This assumption is, however, often far from realistic, as legitimate IC vendors typically restrict access to the scan chain, especially for security-critical use cases [97].

- Due to the assumption that the location of the key inputs is known, the secret key can be leaked by simple design-independent hardware Trojans once the IC is activated, regardless of the underlying locking scheme [82].
- Multiple recent works have demonstrated the successful extraction of keys through probing and fault-injection attacks on locked ICs, even in the presence of a tamper- and read-proof memory [60, 81, 127].

At first glance, these observations exclude the necessity to deploy OG attacks since the key is directly accessible through the *activated* IC, regardless of the locking scheme. However, OG attacks still play an important role in a divide-and-conquer Reverse Engineering (RE)-based scenario, even if a low-volume production is assumed. More details are provided in the next section.

4.8 Logic Locking and Reverse Engineering

In this book, we assume that the attacker is required to perform RE to be able to design and insert controllable, design-dependent hardware Trojans, or in the terminology of Chap. 3, this book focuses on Class-1 hardware Trojans (C1HTs). Hence, the main role of logic locking is to increase the complexity of RE. To understand the impact of logic locking in the RE process, let us consider the flow in Fig. 4.3. In terms of RE, we assume that the attacker has access to the locked flattened netlist, either directly via the untrusted, external design house or by extracting the netlist from the layout. To inject a hardware Trojan, the attacker has to retrieve the exact design details, such as the design's functionality and structural composition. Thus, we assume that the RE process involves the following steps. First, the flattened netlist is partitioned into its subcomponents (Fig. 4.3a), which we refer to as modules. This can be done via clustering algorithms, data path analysis,

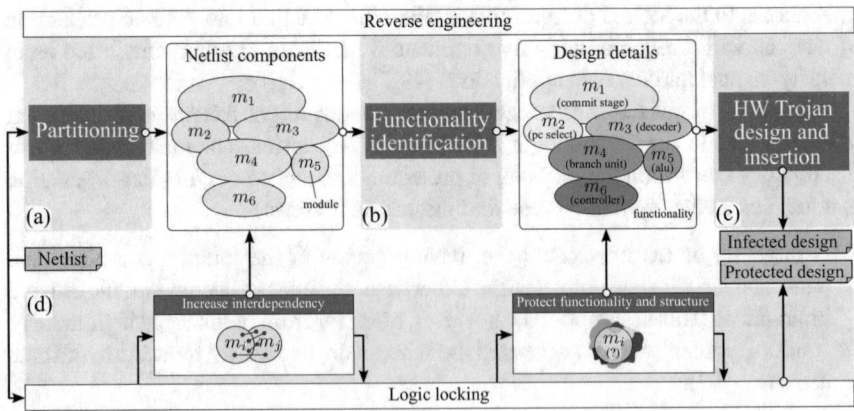

Fig. 4.3 Logic locking and reverse engineering

and other techniques [22, 198]. After the netlist modules have been structurally separated, the next step involves the functionality identification (Fig. 4.3b). The goal of this step is to identify the correct functionality of each module. Assuming that perfect partitioning is performed and a golden model exists, the identification can be done through a formal equivalence check by cross-referencing each partitioned module with a database of designs. Since this is typically difficult to perform due to inherent errors in the RE process, more approximate (fuzzy) methods can be deployed. After functional and structural design details have been uncovered, the attacker can proceed with the design and insertion of a design-dependent hardware Trojan (Fig. 4.3c).

From the introduced procedure, we can identify two focal points that can impact the complexity of RE: the prevention of partitioning and identification. Therefore, as presented in Fig. 4.3d, logic locking must impact the RE effort by (1) increasing the functional and structural interdependency between design modules to prevent partitioning and (2) injecting sufficient design alterations to hinder a successful identification of the design's original functional and structural details. Note that (1) has not been addressed yet in the literature, as existing work typically operates in the OG model, which assumes that the functionality is known (as an activated oracle is available). Nevertheless, even in an OL model, if the functionality is identified, the attacker could gain access to an oracle by referencing a golden model from the database or by assembling an oracle manually. Therefore, it is crucial to address both (1) and (2) to offer sufficient protection against OL attacks and avoid enabling an OG scenario. Moreover, as logic locking and reverse engineering are both elaborate procedures, it still remains a challenge to precisely measure the complexity of RE as well as the security ramifications of logic locking.

4.9 Attack Scenario

As introduced in Chap. 1, the main objective of this book evolves around the development and application of security methodologies to protect the *integrity of hardware designs* against malicious design modifications. Thereby, the presented work builds upon the protection concept provided by logic locking. Thus, in the context of logic locking, the attack scenario considered in this book includes the following assumptions:

- **Attacker:** The external design house, foundry, and assembly facility are considered untrusted. Henceforth, these are collectively referred to as *the attacker*.
- **Attack objective:** The main objective of the attacker is to insert an intelligible, controllable, and design-dependent (class-1) hardware Trojan into a given design. Henceforth, the design under attack is referred to as *the target design*.
- **Attack requirements:**

- To construct and insert the design-dependent hardware Trojan, the attacker requires a sound understanding of the functionality and structure of the target design. Thus, reverse engineering the design is required.
- To successfully perform reverse engineering, the attacker has to unlock the design.

- **Attack type:** The payload of the hardware Trojan is assumed to execute a Denial of Service (DoS) attack. Even though the work presented in this book is applicable regardless of the exact attack the HT performs, we consider that a denial of service embodies the most destructive attack form—specifically if placed in the context of automotive, medical, or defense applications.
- **Target design:** The main target design is considered to be a processor core. The reasoning is as follows. Placed at the heart of every system, a processor core has a direct relation to the executed software as well as to other hardware components, such as peripherals, memories, co-processors, and others. Therefore, we assume that a high-impact hardware Trojan would most likely be placed inside the processor.
- **Design environment:** The target design is placed in a low-volume production setting for security-critical applications. Thus, the attacker has only access to the locked design in gate-level netlist format, either directly as a design house or by extracting the netlist from the layout. Therefore, we concentrate on the OL attack model.

Other supply chain vulnerabilities, such as compromised third-party IPs, untrusted design tools, and in-the-field attacks, are beyond the scope of this book.

4.10 Synopsis

This chapter discussed the basic mechanics of logic locking. First, the standard notation used throughout the book has been introduced. Second, based on a simple example of locking, the security ramifications of locking policies w.r.t. the IC supply chain have been analyzed. Next, the classification of attack models has been introduced alongside an analysis of the interplay between RE and the existing attack assumptions. Finally, the details of the attack scenario used in this book were introduced.

Chapter 5
Attacks and Schemes

More than a decade of research has been invested in the exploration of logic locking from both the defender's and the attacker's point of view. This process has resulted in a wide range of logic locking families that are specialized for the protection against specific attacks. Furthermore, multiple attack types have emerged to tackle different security aspects of logic locking.

This chapter surveys the related work by focusing on the evolution of logic locking since its inception. The evolution of attacks and logic locking schemes is presented in Sects. 5.1 and 5.2, respectively, thereby following Figs. 5.1 and 5.2. The attacks and schemes are discussed in chronological order.[1] The lessons learned are summarized in Sect. 5.3. More information on both schemes and attacks can be found in [80, 217, 218, 223, 224]. Note that this chapter first discusses the evolution of attacks before introducing existing locking schemes. The reason is that many schemes have been developed only to thwart specific attacks. Thus, the attack landscape is introduced first to facilitate further discussions. Note that the presented data focuses exclusively on *combinational logic locking*.

5.1 Evolution of Attacks

With every new iteration of schemes, novel vulnerabilities are detected and exploited to either extract the correct key or remove the locking circuitry. This incremental game has led to novel observations and successive improvements in the security of logic locking. In the following, we present a brief overview of the evolution of attack vectors on logic locking, thereby following the visualization in

---

[1] The order is assembled based on the publication date of the discussed work. Note that some schemes or attacks have been introduced in multiple publications, e.g., papers, journals, or open-access materials. Where applicable, we only consider the first publication date.

© The Author(s), under exclusive license to Springer Nature Switzerland AG 2023
D. Sisejkovic, R. Leupers, *Logic Locking*,
https://doi.org/10.1007/978-3-031-19123-7_5

Figs. 5.1 and 5.2. *Note that the presented data is valid only up to the time of writing this book.*

5.1.1 Classification of Attacks

We introduce a system to classify attacks based on three criteria:

- **Exploitation characteristics:** The exploitation characteristics describe what aspect of logic locking is exploited by an attack.
- **Attack model type:** This criterion defines which attack model is assumed.
- **Result type:** This type describes the format of the attack result.

5.1.1.1 Exploitation Characteristics

Based on what characteristics of logic locking are exploited, we broadly classify attacks into four categories as follows:

- **Functional:** These attacks exploit the various effects that logic locking has on the functional characteristics of locked designs to extract the correct key.
- **Side-channel:** This class exploits key-related information leakage that hides in the form of power, timing, delay, and other design characteristics.
- **Structural:** Structural attacks focus on the exploitation of key- or scheme-related structural residue in the locked design. These attacks can either extract the correct activation key or perform a structural removal of the locking circuitry.
- **Physical:** Attacks in this class exploit physical characteristics of the locked design to acquire knowledge about the key. Typically, physical attacks are in the form of optical probing, fault injection and analysis, Hardware (HW) tampering, and others.

5.1.1.2 Attack Model Type

Each attack class can further be separated into the Oracle-Guided (OG) and Oracle-Less (OL) category as discussed in Sect. 4.7.

5.1.1.3 Result Type

We describe the result type using three distinct features:

- **Removal vs. key extraction:** An attack can result in either the removal of the locking circuitry or the extraction of the key. Even though often these results are equivalent, in some cases the removal does not uncover the correct key.

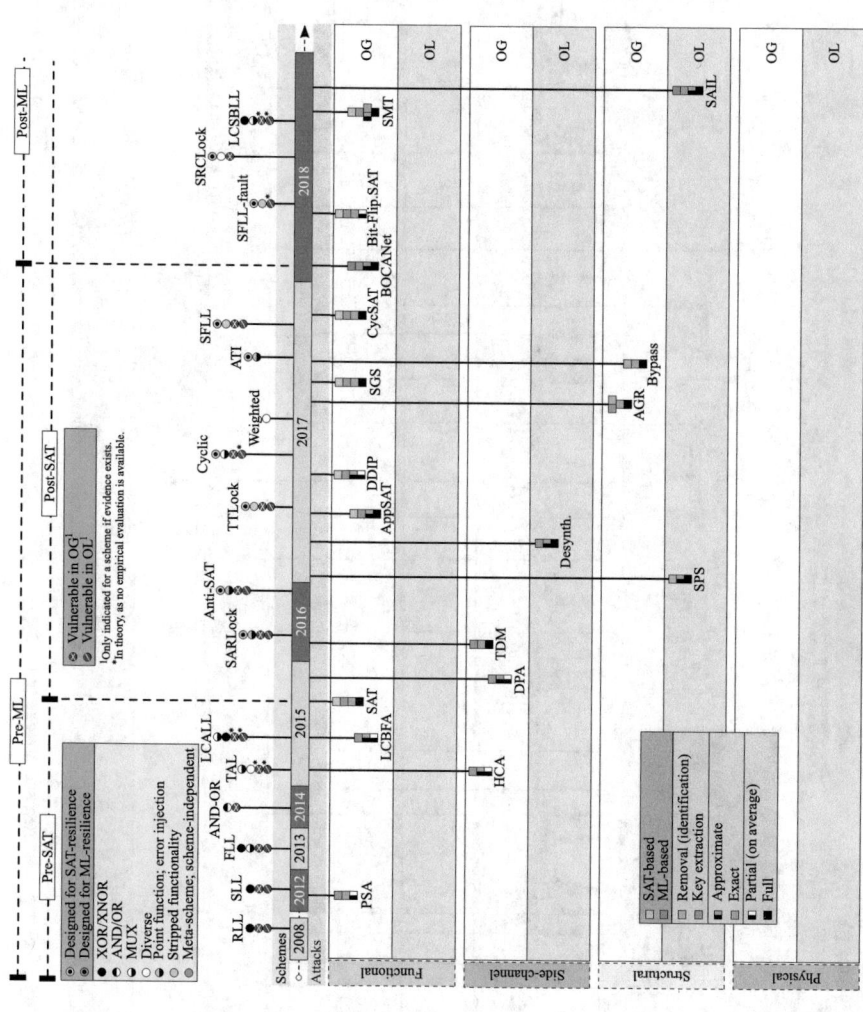

Fig. 5.1 Evolution of logic locking: schemes and attacks (part 1)

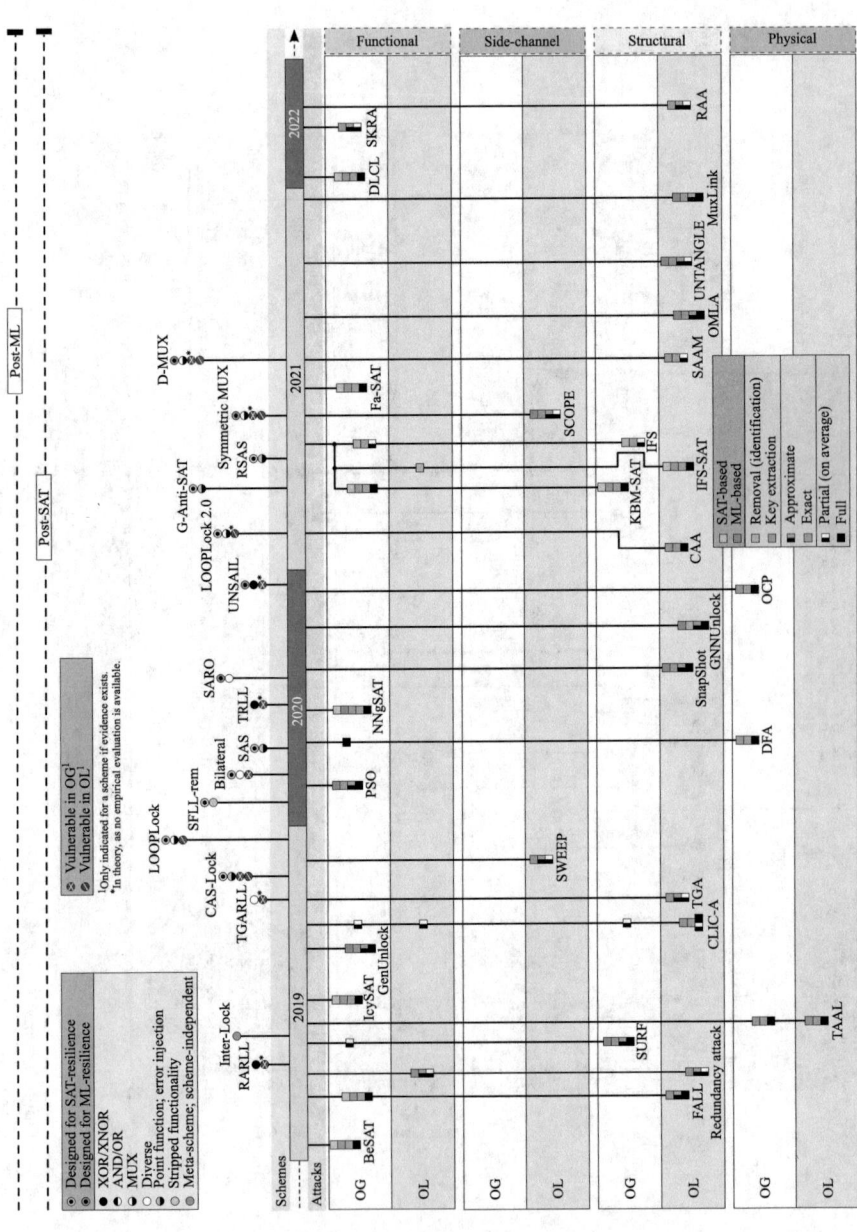

Fig. 5.2 Evolution of logic locking: schemes and attacks (part 2)

- **Approximate vs. exact:** An approximate result indicates that the removal or the retrieved key is not guaranteed to be correct, i.e., *no exact (and full) certainty* is provided by the attack about the correctness of the result.[2] On the contrary, an exact result indicates that the attack can provide a guarantee of the result's correctness.
- **Partial vs. full:** A partial result defines that the result is only partially returned, e.g., only some bits of the key are retrieved.[3] A full result defines a fully retrieved key or full removal of the locking circuitry.[4]

The introduced classification allows for the description of all result types.

5.1.2 Functional Attacks

One of the first functional attacks on logic locking is the Path-Sensitization Attack (PSA) [129]. This OG attack utilizes the test principle of sensitization. By analyzing the structural composition of the locked netlists, PSA computes input patterns that can sensitize individual key bits to primary outputs in the activated oracle, hence making them observable.

The Logic Cone-Based Brute-Force Attack (LCBFA) can be combined with PSA to extract further key bits [98]. This OG attack performs a brute-force search of the key space for each input cone individually, thereby processing the input cones according to the number of containing key inputs in ascending order.

The Boolean Satisfiability Problem (SAT) attack played a crucial role in designing the landscape of logic locking [175]. This OG attack relies on the notion of distinguishing input patterns (DIPs)[5]—patterns that for two different key values generate two different outputs of the locked circuit. In an iterative process, the attack constructs a miter circuit to compare the output of two duplicated circuits. In each iteration, a SAT solver is invoked to find a satisfying assignment for the miter that yields a Distinguishing Input Pattern (DIP). In all successive steps of generating a new DIP, the two input keys of the miter circuit must produce the same output for all previous DIPs as well. This process continues until the SAT solver cannot find a new DIP. Once this point is reached, any key assignment that generates the correct output for the set of found DIPs is the correct key. Based on this efficient procedure of ruling out incorrect keys, a wide range of OG SAT-based attacks have been introduced in the literature, thereby targeting various attack objectives. For example, the AppSAT [151] and Double DIP (DDIP) attack [160] generate

---

[2] An approximate result can also be defined as a result that returns an approximate netlist, e.g., in the form of a partial key [218]. However, we adjust this definition as some attacks can be certain of the result yet return a partial key.

[3] We classify an attack with a partial result if the attack *on average* returns a partial result. Thus, it is possible for a partial attack to return a full key. However, the available empirical evaluations often showcase that partial attacks do not return a full result.

[4] If an attack is classified as returning a full result, its result is *always* full.

[5] Also referred to as discriminating input patterns.

approximate results. The Sensitization-Guided SAT (SGS) attack [217] targets AND-Tree Insertion (ATI) locking by combining a sensitization procedure with the SAT attack. CycSAT [228], Behavioral SAT (BeSAT) [158], and IcySAT [155] target cyclic-based locking policies. The Bit-Flipping SAT attack [159] is designed to compromise compound locking. A generalization of the SAT attack comes in the form of the Satisfiability Modulo Theory (SMT) attack [18]. The NNgSAT attack deploys a neural network to accelerate the SAT attack on SAT-hard circuit structures [19]. The recent Fa-SAT attack integrates a fault-injection procedure to assist SAT-based attacks in converging toward a key faster [105].

One of the first Machine Learning (ML)-based attacks is known as BOCANet [184, 185]. This OG attack leverages recurrent neural networks to learn the correct functional I/O mapping of a locked circuit, thereby providing the subsequent refinement of the activation key based on the learned function. The approximate OG GenUnlock [40] and Particle Swarm Optimization (PSO) attack [88] leverage evolutionary computation to perform a heuristic key search that is steered by selected fitness functions.

The Characterization of Locked Integrated Circuits via ATPG (CLIC-A) attack [55, 56] utilizes common testing procedures to uncover key values, thereby being effective across a large variety of locking schemes. CLIC-A includes multiple attack models, spanning across key-input sensitization, constraint-based ATPG, exploitation of key-dependent faults, and sequential key extraction [80].

The recent DIP Learning on CAS-Lock Attack (DLCL) further compromises CAS-Lock by revealing not only the correct key but also the exact structural implementation of the scheme [142].

An attack on compound locking comes in the form of the Statistical Key Recovery Attack (SKRA) [91]. This OG attack links the key recovery problem with distinguishing two output distributions and deploys Welch's t-test to recover a set of the key bits from a compound locking scheme.

5.1.3 Side-Channel Attacks

The Hill-Climbing Attack (HCA) performs an iterative search of the key space by toggling key bits and observing the differences in the output responses of the locked netlist and activated Integrated Circuit (IC) (or test patterns) [124]. Thus, this OG attack exploits the behavioral side-channel for incorrect keys to extract a correct key value.

The OG Differential Power Analysis (DPA) attack exploits the correlation between the power consumption of the locked circuit and the correct key bits [214]. By using the correct key as a constraint in Automatic Test Pattern Generation (ATPG) for post-production chip testing, the Test-Data Mining (TDM) attack [220] attempts to reverse the ATPG procedure to find a correct key that maximizes the fault coverage.

The first OL side-channel attack is known as the desynthesis attack [112]. This attack successively searches for a key value that minimizes the dissimilarity between the locked netlist and synthesized netlist with a fixed key value. A similar concept is utilized in the OL SWEEP attack [4]. This constant propagation attack exploits the differentiating circuit characteristics induced by hard-coding key-bit values during synthesis. Hence, SWEEP is able to deduce a correlational between the synthesis features and the correct key value. A more advanced version of SWEEP hides within SCOPE [5]. Even though the SCOPE attack is based on similar mechanics as SWEEP, it functions in an unsupervised manner, hence not requiring any training data.

5.1.4 Structural Attacks

One of the first OL removal attacks is manifested in the structural Signal Probability Skew (SPS) attack. SPS utilizes the structural traces induced by the Anti-SAT scheme to identify, isolate, and remove the locking circuitry [216, 217]. The AppSAT-Guided Removal (AGR) further extends this concept to compound locking, specifically to the combination of Anti-SAT and traditional logic locking [217]. AGR integrates AppSAT alongside a structural analysis of the locked netlist to recover the exact target netlist. A different attack approach is taken by the bypass attack [207]. Its methodology is to first identify the DIPs that generate incorrect outputs and afterward mask the erroneous behavior through the insertion of a bypass circuitry.

The first OL ML-based structural attack is known as SAIL [36]. It deploys ML algorithms to reconstruct key-related local logic structures that have been affected by logic synthesis. Hence, SAIL attempts to reverse the effects of the synthesis process and extract the key value based on the structural properties of XOR/XNOR-based locking. SAIL has been further improved in the OG model via the SURF attack [37], by further refining the activation key by means of an oracle-guided structure-aware greedy optimization algorithm.

An important role in the SAT-driven research domain is played by the FALL attack [162, 163]. This attack targets Stripped-Functionality Logic Locking (SFLL)-locked circuits by deploying both structural and functional analyses to detect the locking circuitry and retrieve a correct key. The significance of FALL lies in being one of the first attacks to compromise SFLL—one of the most effective SAT-resilient schemes.

The redundancy attack [99] recovers the key by pruning out incorrect key values that introduce a significant level of functional redundancy in the netlist. Similarly, the Topology-Guided Attack (TGA) [226] recovers correct key values for selected key gates by looking at equivalent functions that are constructed using all possible hypothesis key values. Hence, TGA exploits the repeating nature of basic functions within a netlist.

A notable ML-based attack is known as GNNUnlock [10]. This OL attack performs a removal of SAT-resilient locking circuitry by analyzing the structural characteristics of the netlist gates with a Graph Neural Network (GNN). A similar attack is known as OMLA [13]. This OL attack utilizes GNNs to resolve key-bit values by formulating the key prediction as a subgraph classification problem. Another potent ML-driven attack comes in the form of UNTANGLE [9]. This attack targets Multiplexer (MUX)-based SAT-hard locking by addressing a link prediction problem with GNNs.

The OL Cycle Analysis Attack (CAA) is designed to compromise cyclic locking through a careful structural analysis of cycle pairs within a netlist [211]. In particular, it was designed to attack the LOOPLock scheme [41].

The Identify Flip Signal (IFS), Key-Bit Mapping (KBM)-SAT, and IFS-SAT attack [145] target the (M)-CAS-Lock scheme by exploiting a set of scheme-specific structural vulnerabilities that can successfully compromise the scheme.

The recent OL ML-based MuxLink [12] attack targets MUX-based locking techniques by leveraging GNNs for link prediction. Therefore, MuxLink can predict a correct link through a multiplexer with a high prediction accuracy.

The OL Rationality Analysis Attack (RAA) targets XOR- and MUX-based locking by analyzing the redundancy level deviation for different keys [101]. The basis of this attack relies on the insight that incorrect keys often produce circuits that violate universal design principles.

5.1.5 Physical Attacks

A powerful physical attack is embodied in the Tampering Attack on Any Key-Based Logic Locked Circuit (TAAL) [82]. Compared to other attacks, TAAL takes an invasive approach to break a scheme, by implanting stealthy hardware Trojans in the target netlist that leak the secret key to the attacker once the chip is activated. As the key inputs are always identifiable according to the common attack assumptions (see Sect. 4.7), TAAL can compromise logic locking without the necessity to attack the scheme itself.

The OG Differential Fault Analysis (DFA) attack [81] demonstrates key extraction by injecting faults[6] in the key lines and comparing the output response with the fault-free counterpart.

Finally, the Optical Contactless Probing (OCP) attack [127] demonstrates a successful extraction of key values from the activated chip via electro-optical frequency mapping even in the presence of tamper- and read-proof memory.

---

[6] Fault injection equipment is required to launch the attack. Hence we categorize it as a physical attack.

5.2 Evolution of Schemes

The following introduces the landscape of logic locking schemes, thereby building on the already discussed attacks from the previous section.

5.2.1 Classification of Schemes

We classify logic locking schemes based on the main building blocks of their implementation. These include XOR/XNOR gates, AND–OR gates, MUXs, diverse gates, point functions (error injection blocks), stripped-functionality-based implementations, and meta-schemes. Moreover, the evolution of logic locking has been steered by the introduction of novel attacks, resulting in the definition of new security objectives. A notable attack is the SAT attack that was introduced in 2015 [175] (see Sect. 5.1). This OG attack was capable of breaking all known locking schemes at that time. Hence, the developmental timeline of logic locking is often partitioned into pre- and post-SAT schemes.[7] Moreover, ML-based attacks have become more prominent in recent years. Hence, we split the timeline into the pre- and post-ML age.

5.2.2 Pre-SAT Schemes

The first logic locking scheme is known as Random Logic Locking (RLL) [140, 141]. This scheme relies on the random distribution of XOR/XNOR key gates in the netlist.[8] As presented in Sect. 4.2 and visualized in Fig. 5.3a, if the correct key bits are provided to the inserted key gates, the locked netlist remains functionally equivalent to the original, i.e., the non-key inputs of the key gates are buffered. The main security objective of RLL is to ensure the dependency of the correct circuit behavior on the activation key.

Strong (Secure) Logic Locking (SLL) was later introduced based on the same locking principle [129, 219]. However, the insertion methodology of SLL targets the prevention of the PSA [129] attack by ensuring a high functional interference among the inserted key gates.

The random distribution of XOR/XNOR key gates might not always ensure a high output corruptibility. This security property is typically measured in terms of the Hamming Distance (HD) between the correct and corrupted circuit responses. A

---

[7] Note that many post-SAT schemes are not focusing on SAT resilience. Nevertheless, we can place them in the post-SAT era. In the following sections, we indicate if SAT resilience is the main objective of a scheme.

[8] RLL was initially introduced to support IC metering. Later, it was categorized as logic locking.

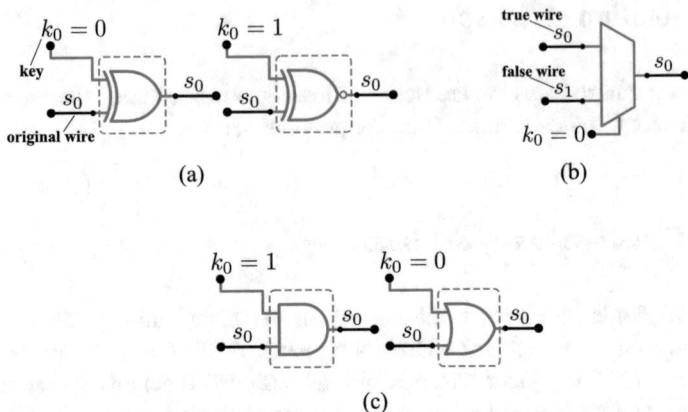

Fig. 5.3 Working principles of pre-SAT logic locking: (**a**) XOR/XNOR, (**b**) MUX-based, and (**c**) AND/OR

50% HD value maximizes the ambiguity of the output behavior for incorrect keys. To improve this security aspect, Fault Analysis-Based Logic Locking (FLL) inserts XOR/XNOR gates or MUXs in locations that ensure the highest impact on the circuit outputs for incorrect keys [132]. In MUX-based logic locking, the inserted MUXs are driven by a true (correct) and false (incorrect) wire, as presented in Fig. 5.3b. If the correct key bit is applied, the true wire is buffered through the MUX.

The AND–OR locking scheme aims at minimizing the number of low-controllability signals (referred to as rare values) in a netlist to prevent Hardware Trojan (HT) insertion [53]. The assumption of this scheme is that an attacker is likely to bind HTs to signals with low controllability in order to achieve a stealthy Trojan implementation. To mask these locations, the AND-OR scheme inserts key-controlled AND OR gates that increase the probability of 0 or 1 values according to the controllability measure. In terms of functionality, AND OR gates can act as buffers or inverters similar to XOR/XNOR gates, as presented in Fig. 5.3c.

The Test-Aware Locking (TAL) scheme tackles the problem of pre-activation testing [124]. Fabricated ICs must be unlocked prior to functional testing. If this task is outsourced to an untrusted party, the activation key remains exposed to attacks, including the HCA, as discussed in Sect. 5.1. To overcome this challenge, TAL deploys a MUX-based locking procedure that allows for the preservation of test responses without exposing the key.

Logic Cone Analysis Logic Locking (LCALL) tackles the problem of brute-force attacks [98]. The number of brute-force key combinations can be reduced by localizing all key gates that impact each primary output individually. This can be performed via LCBFA [98]. By intelligently deploying MUXs, LCALL increases the number of output-impacting key gates as well as the complexity of a brute-force attack per primary output.

Note that most pre-SAT schemes have been successfully attacked with the SAT attack in the OG attack model. Moreover, TAL and LCALL include schemes that are most likely vulnerable to the SAT attack and Structural Analysis Attack on MUX-Based Locking (SAAM) [166].

5.2.3   Post-SAT Schemes

The SAT attack has left a deep mark on the landscape of logic locking in the OG attack model. Since its introduction in 2015, many efforts have been invested by the security community to design SAT-resilient schemes. These efforts have evolved around three focal points: Point functions (PFs), SAT-unresolvable structures, and stripped-functionality logic locking.

5.2.3.1   Point Function-Based Schemes

PFs are Boolean functions that exhibit a very specific behavior that thwarts the SAT attack: the value 1 is produced only for one input pattern. Examples are AND NOR functions. The concept of PFs can be extended to support key inputs, i.e., in the form of a controllable error-injecting locking policy. Naturally, PFs result in an undesirable security implication: a very low output corruptibility. Hence, it has become a major challenge to balance SAT resilience and output corruptibility, as discussed in Sect. 6.2.1.

SARLock was introduced in 2016 as one of the first PF-based schemes [215]. The scheme is constructed as shown in Fig. 5.4a. For each incorrect key, an error is injected by flipping a selected primary output signal. The flip occurs only for one selected input pattern per key. This PF-based behavior forces the initial SAT attack to rule out only one incorrect key per iteration, hence collapsing to a brute-force attack.

A similar PF-based scheme is known as Anti-SAT [204, 206]. This scheme implements two complementary functional blocks g and $\overline{g}$, as presented in Fig. 5.4b. Both blocks receive the same inputs (a set of selected circuit wires X) and a separate set of keys. The outputs of g and $\overline{g}$ are fed into a 2-input AND gate, generating the output Y. Hence, $Y = g(X, k_0, \cdots, k_n) \wedge \overline{g(X, k_{n+1}, \cdots, k_{2n})}$. Y is injected into the original circuit via an XOR gate. Therefore, for an incorrect key, the XOR gate inverts a selected signal, thereby corrupting the behavior of the circuit.[9] Both SARLock and Anti-SAT have been successfully attacked through a combination of SAT-based and removal attacks [159, 216, 217, 224]. Moreover, Anti-SAT is vulnerable to the OL ML-based attack GNNUnlock [10].

---

[9] An AND-based Y forms a type-0 Anti-SAT. Similarly, an OR-based Y forms a type-1 Anti-SAT. The type indicates if a 0 or 1 is always generated for a correct key. In case of type-1, Y is injected via an XNOR gate.

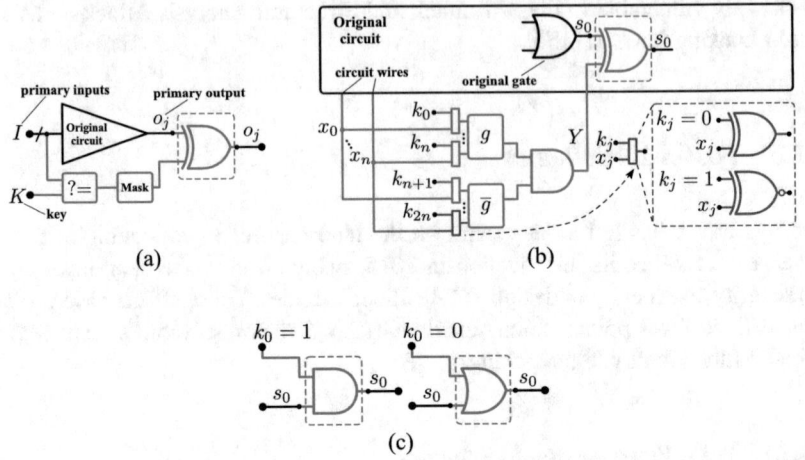

Fig. 5.4 Working principles of post-SAT logic locking: (**a**) SARLock, (**b**) Anti-SAT, and (**c**) TTLock ($h = 0$)/SFLL-HD

The ATI scheme takes a slightly different approach compared to Anti-SAT [102]. Instead of integrating an additional (external) PF construct, ATI searches for suitable AND or trees within the original circuit, hence reducing the implementation cost. Once identified, the tree inputs are locked with XOR/XNOR key gates. Nonetheless, the SGS attack was able to break ATI [217].

An attempt to achieve SAT resilience alongside non-trivial output corruptibility is manifested in the form of (M)-CAS-Lock [150]. To exhibit this behavior, the scheme deploys a cascade of key-controlled AND/OR gates that converge into a single AND gate whose output is used to inject errors into the original circuit. However, the specific structural traces of this scheme enable a range of attacks, including IFS, KBM-SAT, and IFS-SAT [145].

The concept of Anti-SAT was further enhanced in the form of Strong Anti-SAT (SAS) [107], Robust SAS (RSAS) [108], and (G-)Anti-SAT [230]. SAS and RSAS extend Anti-SAT through a function block to simultaneously enable high output corruptibility and SAT resilience. (G-)Anti-SAT represents a generalization of the Anti-SAT scheme by enabling greater flexibility in designing the functions that are integrated into the Anti-SAT block.

5.2.3.2   SAT-Unresolvable Structures

Another potential way to thwart the SAT attack is to distribute SAT-unresolvable structures within the hardware design. One such scheme is known as cyclic logic locking [152]. This scheme inserts MUXs to create non-reducible combinational cycles with the goal to disable the SAT attack, since it was thought to be dependent on acyclic circuit representations. However, this cyclic locking scheme has been successfully challenged by the CycSAT attack [228]. As an improvement to cyclic locking, LOOPLock [41] and SRCLock [138] were later introduced to defend against removal attacks and CycSAT. Nevertheless, MUX-based cyclic locking schemes are potentially vulnerable to SAAM [166] and have all been successfully challenged by multiple advanced variants of the SAT attack [18, 158]. Moreover, LOOPLock can be broken via CAA [211]. A mitigation to CAA was presented in form of the enhanced LOOPLock 2.0 scheme [211].

5.2.3.3   Stripped-Functionality Schemes

A powerful SAT-resilient scheme is manifested in the form of SFLL. As visualized in Fig. 5.4c, SFLL-based schemes strip part of the functionality from the original circuit and reassemble it via the restore unit. This unit guarantees the preservation of the original functionality for a correct key as well as SAT-resilient behavior for a selected set of protected input patterns for incorrect keys. Moreover, the restore unit can be configured to protect $\binom{|K|}{h}$ input patterns that are of Hamming distance h from the $|K|$-bit key. Based on this concept, multiple stripped-functionality schemes have been proposed, including Tenacious and Traceless Logic Locking (TTLock) [222], SFLL-HD/flex [221], SFLL-fault [147], and SFLL-rem [146]. Nevertheless, novel attacks have been introduced to challenge various aspects of selected SFLL variants [10, 14, 162, 163, 209].

Another scheme that can be built on top of SFLL is known as bilateral locking [137]. By applying a combination of locking and routing obfuscation to sensitive hardware components, bilateral locking tries to resolve two common issues of SAT-resilient schemes: low output corruptibility and structural isolation of the locking circuitry. Bilateral locking has been successfully attacked with Fa-SAT [105].

5.2.3.4   SAT-Oblivious and Compound Schemes

A variety of schemes in the post-SAT era do not directly focus on thwarting the SAT attack but rather address other security properties that are orthogonal to SAT resilience.

Weighted logic locking implements a methodology that offers both high output corruptibility and resilience against key sensitization [89]. This is achieved by controlling each key gate with multiple key inputs, as opposed to using a single key input per key gate. Logic Cone Size-Based Logic Locking (LCSBLL) concentrates on placing keys in the largest logic cones to maximize the impact on other signals [16]. Redundancy Attack Resistant Logic Locking (RARLL) inserts key gates with the aim to prevent redundancy attacks [100]. TGA-Resistant Logic Locking (TGARLL) prevents topology-guided attacks by avoiding repeated locked constructs in the circuit [226]. Truly Random Logic Locking (TRLL) enhances the concept of RLL by ensuring the existence of equally likely XOR/XOR-INV gates driven by both

0 and 1 keys [104], thereby disabling structural key-guessing attacks enabled by predictable synthesis transformations.

Without additional security mechanisms, LCSBLL, RARLL, TGARLL, and TRLL are in theory vulnerable to the SAT attack.[10] A compound scheme can be deployed to ensure SAT resilience alongside other security properties. This scheme type typically combines pre- and post-SAT schemes to exhibit both SAT resilience and higher output corruptibility. For example, a popular compound scheme is the combination of SARLock and RLL.

5.2.4   Post-ML Schemes

The introduction of efficient and easy-to-use ML models has started to impact the design space of logic locking. Thus far, only a few schemes have been developed to target ML resilience. More details can be found in [170].

A new model of locking is manifested in the form of Scalable Attack-Resistant Obfuscation (SARO) [3]. This scheme operates in two steps. First, SARO partitions the target design into multiple components to maximize the structural locking-induced alterations. Second, a key-based systematic truth table transformation is deployed to change the functionality of each partition. Since the transformation maximizes the structural and functional randomness of the target circuit, SARO might be efficient in protecting against ML-based attacks.

The UNSAIL scheme has been developed to thwart attacks that target the resolution of structural transformations induced by logic synthesis [11]. The core mechanism lies in the generation of confusing data that leads to false predictions, hence thwarting ML-based attacks. Note that UNSAIL is not primarily designed to withstand SAT-based attacks.

The symmetric MUX-based locking scheme [5] carefully inserts multiplexers to eliminate the structural asymmetry utilized by the SCOPE attack. Note that this scheme can be considered a special case of Deceptive Multiplexer Logic Locking (D-MUX) (discussed in Chap. 10).

5.2.5   New Directions in Logic Locking

5.2.5.1   Routing-Based Locking

A promising approach to thwarting structural and SAT-based attacks is known as routing-based obfuscation. The core mechanism of this locking type includes the

---

[10] An empirical evaluation is not available, and however, the schemes are based on locking principles that are vulnerable to SAT-based attacks.

construction of symmetric interconnections in the circuit to increase the SAT-solving time.[11] Routing-based schemes are built on top of key-programmable routing blocks, which can be assembled into a variety of topologies, including crossbar or permutation networks [86]. Notable routing-based schemes are Cross-Lock [153], Full-Lock [85], Banyan locking [177], and Inter-Lock [86].[12] Note that Cross-Lock and Full-Lock have been broken by the canonical prune-and-SAT attack [86]. Inter-Lock has been successfully challenged by UNTANGLE [9]. Moreover, this group of locking schemes often suffers from a higher area and delay overhead.

5.2.5.2 LUT-Based Locking

Another promising implementation of logic locking lies in utilizing lookup tables (LUTs) [24, 110]. By replacing part of the target design with generic lookup tables that are configured after fabrication, this locking type can potentially resolve any attack vectors in the OL model. Nevertheless, deploying Lookup tables (LUTs) comes at a higher implementation cost and often requires the reliance on novel technologies, hence putting another burden on the traditional hardware design flow. Note that some of the routing-based schemes utilize LUTs to construct the routing block-based topologies.

5.2.5.3 Parametric Locking

Another area of logic locking concerns the problem of parametric locking [222]. This locking type strives to protect the parametric behavior of a design, including power, delay, and reliability. Only a correct activation key ensures the predetermined parametric traits, such as high performance or low power. Compared to traditional logic locking, parametric locking does not necessarily induce incorrect output behavior for incorrect keys.

An active field of research lies within delay locking. This locking type inserts delay keys that create a dependency between the delay profile of the synthesized design and the key value [35]. Therefore, an incorrect key can lead to timing violations and, consequently, a malfunction of the circuit [205].

5.2.5.4 HLS and RTL Locking

Recent efforts have focused on the application of logic locking concepts on higher abstraction levels [78, 79, 87, 121–123]. By deploying locking mechanisms on

---

[11] The SAT attack can be mitigated either by forcing an exponential number of SAT iterations or by increasing the time of solving each iteration individually.

[12] Not to be confused with the Inter-Lock scheme presented in Chap. 8.

Register-Transfer Level (RTL) designs or during High-Level Synthesis (HLS), these locking policies perform semantic manipulations of the technology-independent, high-level design representation. The assumed advantage of HLS/RTL locking includes having access to all semantic information of the hardware design, whereas gate-level locking operates after logic synthesis, which can potentially absorb critical information (such as constants) [121]. Furthermore, state-of-the-art RTL locking has recently been evaluated in the context of machine learning-based attacks [165].

5.2.5.5   Universal Circuits

A powerful security concept hides within universal circuits [156, 229]. These circuit types are capable of implementing any function in a selected circuit family by adjusting the configurable inputs, hence replicating the concept of field-programmable gate arrays (FPGAs) embedded in an Application-Specific Integrated Circuit (ASIC). The security of universal circuits is provided by the fact that their structural composition is not dependent on the functionality of the target hardware design. Nevertheless, the universality comes with impractical cost implications.

5.3   Lessons Learned

The vast landscape of attacks makes it increasingly difficult to design resilient schemes. As can be seen in Figs. 5.1 and 5.2, most locking schemes have been compromised in at least one attack model.

Moreover, the novel class of physical attacks is capable of unlocking any scheme in the OG model, thereby overshadowing the necessity for a range of effective non-physical OG attacks. The main reason for the success of physical attacks is the fact that the security of the key is not guaranteed once the chip is produced, hence making it possible to fully circumvent the security properties of the underlying scheme. Note that it has long been assumed that the key is kept secret in the presence of a read- and tamper-proof memory. Evidently, the security of the stored key is still an open challenge. Nevertheless, logic locking plays a crucial role in securing hardware designs by counteracting the reverse engineering process. Furthermore, the overwhelming variety of security properties that are addressed by different schemes and attacks make the precise measuring of the security of logic locking an extremely challenging task. All these factors indicate an urgent need to rethink the objectives of logic locking. Therefore, in this book, we focus on evaluating the effectiveness of locking policies within the frame of a concrete, concise, and realistic attack model. Based on the attack scenario defined in Sect. 4.9, in this book, we discuss the following:

- Extensible security metrics for logic locking.
- Inter-Lock: a meta-scheme that transcends the security properties of a single locking scheme, thereby focusing on an OL reverse engineering-based attack model on complex hardware designs. To this day, the security concepts of the scheme have not been broken.
- D-MUX: an ML-resilient scheme, designed to provide protection in the OL attack model.
- SnapShot:[13] an ML-based OL attack that challenges the fundamental structural features of logic locking schemes.
- SAAM: a structural OL attack on multiplexer-based logic locking.

5.4 Synopsis

This chapter provided an overview of the developments in logic locking since its inception, thereby introducing a systematic classification system for attacks. Furthermore, the lessons learned summarized the critical points in designing logic locking.

---

[13] SnapShot was officially published in 2021 [169], and however a preprint is available since 2020 [168].

Chapter 6
Security Metrics: One Problem, Many Dimensions

Despite the great efforts invested in designing resilient schemes, the versatile security properties of logic locking make it notoriously difficult to define usable and comprehensive security metrics. Consequently, it becomes a challenge to evaluate and compare different locking schemes. The lack of a precise notion of security can be traced back to the following reasons. First, the security of existing locking schemes is highly dependent on the data to be locked, both in structure and in functionality. For example, the corrupted functional output for incorrect keys often includes hints to the key itself. Second, the structural changes induced by logic locking remain present in the Integrated Circuit (IC) even after activation is done. Both aspects enable a plethora of attack vectors, making it tremendously difficult to compile all relevant security properties into a single metric. Third, most defenses explicitly thwart only specific attacks, disabling the possibility for a generalized comparison. Finally, the security properties of logic locking tend to fluctuate with a changing attack model. All these aspects have led to security metrics being a stumbling block for logic locking since its inception. Moreover, security and cost have always been at crossroads in many domains. As Hardware (HW) is typically designed through a careful balance of the area, power, and performance implications, not having a security notion makes an adaption of logic locking in an industrial setting impractical.

The mentioned challenges offer some perspective on the design of security metrics. For example, the variable nature of the security of logic locking suggests that a comprehensive metric must include both functional and structural security aspects in a multidimensional model. Using this concept, in this chapter, we introduce a unifying hardware security metric for logic locking evaluation [192]. Hereby, the metric is constructed by compartmentalizing crucial security aspects into a multidimensional set of properties that span across the various characteristics of existing schemes. Moreover, as logic locking is a rapidly evolving field of research, it is impossible to predict how security properties will change in time. Thus, the metric is designed to support future adaptations.

© The Author(s), under exclusive license to Springer Nature Switzerland AG 2023
D. Sisejkovic, R. Leupers, *Logic Locking*,
https://doi.org/10.1007/978-3-031-19123-7_6

Furthermore, the measure of an acceptable locking-induced overhead has long been discussed in the scientific community without a tangible outcome. However, the price tag for security often has a significant impact on the resilience level of a locking policy. Thus, in addition to the metric design, in this chapter, we present a discussion and analysis of the security-cost trade-off using a concrete attack vector [232].

The rest of this chapter is organized as follows. Design objectives and a metric classification system are presented in Sect. 6.1. Sections 6.2 and 6.3 discuss the functional and structural dimensions of the proposed metric. An experimental evaluation of the metric system is presented in Sect. 6.4. The analysis of the security-cost trade-off problem is presented in Sect. 6.5. Limitations are discussed in Sect. 6.6. Related work is presented in Sect. 6.7. Finally, Sect. 6.8 concludes this chapter.

6.1 Dimensions of Security

Let us first discuss the security dimensions of logic locking and how to approximate the overall security of a locked design, guided by Fig. 6.1. As discussed in Sect. 4.5, one of the main objectives of logic locking is to ensure functional secrecy, i.e., to hide the true functionality of a design. However, as this is difficult to express and measure, the security of logic locking is typically approximated through the retrievability of the correct key. We formulate this aspect of a key in terms of the key-space size (more details in Sect. 6.1.1). A locked design is fully broken if a set of attacks is capable of reducing the total key-space size (KSS) of a locked design to zero. The question is: what aspects of a locked design can be attacked to reduce the

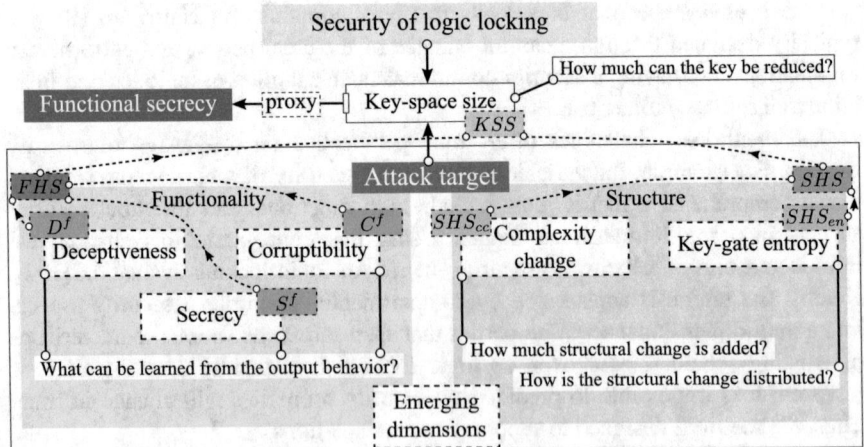

Fig. 6.1 Dimensions of security

key-space size? Logic locking has a twofold impact on a HW design. First, it creates a dependency between the key and the functional behavior of the locked design. Second, it changes the physical (topological) structure of the design through a set of key-dependent transformations. Thus, an attack can target either the functional or structural dimension of logic locking to fully or partially retrieve the correct key.

The exact attack target often depends on the available resource assumptions. For example, in an oracle-guided attack model, the attacker is likely to exploit functional aspects of the locking scheme, e.g., through the behavior of the locked HW for incorrect keys. To capture this security aspect, we define the difficulty of uncovering the activation key through a functional analysis as functional hardware security (FHS). Hereby, FHS covers the security captured in the functional output behavior. This aspect of security is described in Sect. 6.2.[1]

If an oracle is not available, the attacker typically resorts to the analysis of the structure (topology) of the locked design. Hereby, the amount (complexity change) and placement of the key-affected structures (key-gate entropy) offer a path toward compromising a scheme. Therefore, we introduce the structural hardware security (SHS), described in Sect. 6.3.

In the following, the design concept of each metric dimension is to define how near a locked HW component is to a theoretically perfectly secure design. All metric dimensions are scaled to values from 0 to 1, which can be translated to percentages from 0 to 100%.

6.1.1   The Key-Space Size

The efficiency and success of an attack can be limited by the minimum set of keys the attack is forced to run through in a brute-force manner, even if the best heuristic approach and highest computational power are available. We define this security notion as KSS. In a perfectly secure design, the attacker is forced to search the entire key space of size $2^{|K|}$, where Ks is the activation key in the form of a vector of bits. For a sufficiently large K, the probability of retrieving the correct key through a brute-force approach is negligible. Unfortunately, due to the functional nature of logic locking and the target HW, the key space depends not only on the key length but also on the features of the key-gate placement, yielding the total search space to possibly be much less than $2^{|K|}$. Thus, it is necessary to compile a clear measure in the presence of attacks that try to reduce the key space. Inspired by the definition

---

[1] Note that the key-space size was described as part of FHS in [164] as previously most attacks were focusing on the functional aspects of schemes. However, as the landscape of attacks has evolved, extracting the key-space size beyond the functional and structural aspects allows for a more precise measure of security.

of golden patterns,[2] we propose to use the *golden factor*, as a practical notion of security in terms of the KSS:

Definition 6.1 The **golden factor** (G^fs) represents the currently assured key length that implies a lower bound key-space size.

Thus, for a locked design, G^fs indicates that the minimum number of keys an attacker is forced to search through is $KSS = 2^{G^f}$. This key-space size must be assured after all known key-space reduction attacks have been deployed. Hereby, a key-space reduction attack is any attack that is able to rule out parts of the key space. To generalize this model, we introduce the definition of *key-space decrease functions* as follows:

Definition 6.2 A **key-space decrease function** (f^{dc}s) is a function returning a set of key-bit indexes uncovered by a selected attack.

In a functional sense, f^{dc}s takes the IC_{ll} as input and returns the set of uncovered key bits. Since the G^f value for a particular design can be updated with every new attack that decreases the total key space, G^f can be expressed as follows:

$$G^f(IC_{ll}) = |K| - |f_1^{dc}(IC_{ll}) \cup, \ldots, \cup f_n^{dc}(IC_{ll})|$$
$$= |K| - \left| \bigcup_{i=1}^{n} f_i^{dc}(IC_{ll}) \right|. \tag{6.1}$$

Here, the expression $\left| \bigcup_{i=1}^{n} f_i^{dc}(IC_e) \right|$ returns the length of the union of all uncovered key-bit indexes.

Essentially, all known attacks that are capable of retrieving key bits are covered within the definition of f^{dc}s. Note that attacks that remove the locking circuitry are also described by the above definition. Removing the circuitry directly corresponds to reducing the key-space size. More details on possible attacks can be found in Chap. 5.

The design concept of G^fs makes it extensible, since it can be calculated for any algorithmic approach, even for those that are yet to come. This ensures a usable key-space metric, as its value can be updated in case novel attacks are able to lower G^f for a given design. In addition, since $0 \le G^f \le |K|$, the highest security in terms of the key-space size is achieved if G^f equals the total key length $|K|$. In this case, $\left| \bigcup_{i=1}^{n} f_i^{dc}(IC_{ll}) \right| = 0$, i.e., all possible f^{dc}s are not able to retrieve any key bits.

---

[2] Golden patterns are defined within the Fault Analysis-Based Logic Locking (FLL) scheme and the Path-Sensitization Attack (PSA). A golden pattern is able to simultaneously perform muting and sensitization of different key gates. In some key-gate configurations, if a golden pattern does not exist, the attacker is forced to perform a brute-force attack [219].

6.1.2 Design Objectives and Classification

Only a generic metric can be used as a criterion for comparing different locking methodologies. Thus, we strive to draft *complete* security metrics, defined as follows:

Definition 6.3 A **complete logic locking security metric** is independent of all algorithmic locking features and universally applicable at any design level.

The universal applicability can be derived based on a technology-independent representation of HW, such as a generic gate-level netlist. Algorithmic independence, however, is far more challenging, as it implies quantifying security independent of the underlying locking policy. In the following, we present a metric model that fulfills both requirements by introducing multiple security dimensions. These are drafted based on a metric classification system as presented in Fig. 6.1.

6.2 Functional Hardware Security

In order to construct the FHS metric, we place the target design in the oracle-guided scenario, where all input and key values of the design can be set, and the output signals can be observed. The implementation nature of logic locking often leads to specific corrupted output patterns which can be exploited to gain insights about the key. Convergence toward a correct key can be achieved through, e.g., Boolean Satisfiability Problem (SAT)-based attacks or a simple Hill-climbing attack, among others. To cover possible attack vectors in this setting, we introduce three FHS dimensions: deceptiveness, corruptibility, and secrecy.

6.2.1 Functional Deceptiveness

Deceptiveness captures the number of correct outputs for an incorrect key, thus measuring how often incorrect keys falsely claim to be correct. We propose to measure deceptiveness through the *deceptiveness factor*, defined as follows:

Definition 6.4 The **deceptiveness factor** (D^f) measures the level of deceptiveness of a logic-locked design through the output behavior for incorrect keys.

The highest value of D^f is achieved if the output for a selected incorrect key matches the correct output for all inputs except for one. Thus, any incorrect key would, in most cases, falsely suggest being the correct activation key. The described deceptive behavior is crucial in thwarting SAT-based attacks, as their efficiency declines for higher D^fs values. To measure D^f for a selected key K and an input set $\mathcal{I}$, we propose the following expression:

Fig. 6.2 Example of
XOR/XNOR-locked netlist

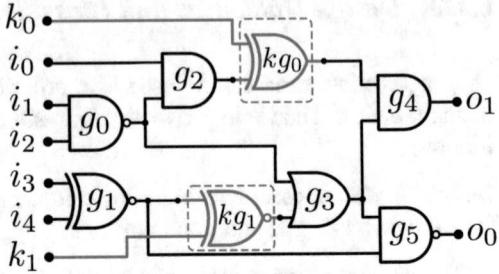

$$D^f = \frac{|O^d_K|}{|\mathcal{I}| - 1},\tag{6.2}$$

where the set O^d_K contains all the *matching* outputs for a selected incorrect key K across all inputs from $\mathcal{I}$s. In this definition, each vector represents an array of bits. For a set of incorrect keys $\mathcal{K}$, the average over all D^f_i is taken, where $1 \leq i \leq |\mathcal{K}|$. Thus, $D^f \in [0, 1]$, with a higher value indicating higher deceptiveness.

Even though high deceptiveness is a key mechanism in thwarting SAT-based attacks, intuitively, a low value of D^f is more desirable as it implies that a randomly selected key is very likely to lead to a completely incorrect output behavior. Thus, a trade-off between D^f and the output corruptibility exists.

6.2.1.1   Example: Deceptiveness Factor

Let us consider the calculation of D^f for the netlist in Fig. 6.2. For a randomly selected key $K = [0, 0]$, the output of both the original and the locked netlist for a set of inputs $\mathcal{I}$ is captured. Since the input space is limited to 2^5 patterns, all inputs are taken into account. Matching outputs are added to the set $O^d_{[0,0]}$. Finally, the size of the accumulated set is $|O^d_{[0,0]}| = 28$, yielding the functional deceptiveness factor $D^f_{[0,0]} \simeq 0.9$. This implies that the selected key results in a correct output for around 90% of inputs.

6.2.2   Functional Corruptibility

The latter can be conveyed into the C^f defined as follows:

$$C^f = \underset{I \in \mathcal{I}, K \in \mathcal{K}}{Prob}[IC_{ll}(I, K) \neq IC(I)],\tag{6.3}$$

where ICs is the original (unlocked) circuit. This metric captures the probability that a locked circuit generates incorrect outputs across all input patterns and incorrect keys. Thus, a direct relation with D^f exists in the form of $C^f = 1 - D^f$. Therefore, a high value of C^f implies low deceptiveness. Striking the right balance between these two factors is still part of active research [150]. Nevertheless, in this work, we propose to look at both factors to convey a comprehensive notion of security.

6.2.3 Functional Secrecy

The final sub-dimension of FHS expresses the secrecy of a locked design through the notion of revealed information for incorrect keys. By adapting Claude Shannon's definition of perfect secrecy [157], we express secrecy in the context of logic locking as follows:

Definition 6.5 **Secrecy** of a logic-locked design is achieved if for any two selected keys $K_1, K_2 \in \mathcal{K}$ and all outputs $O \in \mathcal{O}$s:

$$Prob[IC_{ll}(I, K_1) = O] = Prob[IC_{ll}(I, K_2) = O], \qquad (6.4)$$

where I is randomly chosen from the input space $\mathcal{I}$.

Thus, for any two selected keys and a randomly selected input, the probability that IC_{ll} generates the output Os based on the provided input Is is equal for both selected keys. This means that the attacker is not able to distinguish which key implies which output if given the output for a selected input. In terms of implementation, the maximum secrecy is achieved if $\forall K \in \mathcal{K}$ and $\forall I \in \mathcal{I}$, the output generated by $IC_{ll}(I, K)$ is either a selected constant value or randomly chosen pattern from $\mathcal{O}$s. To measure the secrecy, we introduce the *secrecy factor*, defined as follows:

Definition 6.6 The **secrecy factor** (S^f) measures the level of secrecy of a logic-locked design through the output behavior for incorrect keys.

As for all other discussed factors, the secrecy factor can be manifested in various implementations. In this work, we propose to measure S^f based on a frequency analysis to capture both the constant and random output notion of secrecy. The core idea is as follows. First, all output patterns are recorded for a set of inputs and a selected key. Next, all outputs are stored in a histogram representing the frequency of occurrence of each output value. The highest S^f value is achieved if, for example, all output patterns have the same probability of occurrence for any incorrect key. If this holds true, no information is leaked about the correlation between the key and the output behavior.

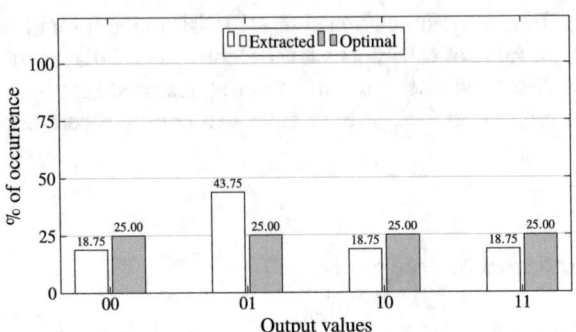

Fig. 6.3 Example of secrecy factor (S^f) calculation

6.2.3.1 Example: Secrecy Factor

The following presents a simple calculation for S^f based on a frequency analysis that can capture both the random and the constant output notion. Let us consider the locked netlist in Fig. 6.2 for the key $K = [0, 0]$. First, the output of the locked netlist is recorded for a set of random input patterns for the selected key. The output patterns are stored in the form of a frequency histogram H, as shown in Fig. 6.3. H captures how often each output value occurs. Considering H, maximum secrecy is achieved if every output pattern is equally likely to appear for a random input and a random key. If this holds, nothing can be learned about the keys from the outputs.

In order to measure S^f, the extracted (v_{ext}) and optimal (v_{opt}) feature vectors are calculated. v_{ext} stores the actual occurrence percentages, while v_{opt} represents the optimal case in which all outputs are equally likely. In this example, the vectors are as follows: $v_{ext} = [18.75, 43.75, 18.75, 18.75]$ and $v_{opt} = [25, 25, 25, 25]$. Finally, S^f is calculated as the Euclidean distance between v_{ext} and v_{opt}, scaled to the maximal distance vector v_{max}. Note that v_{max} represents the case when all inputs map to the same output patter, i.e., $v_{max} = [100, 0, 0, 0]$. Thus, the resulting secrecy is $S^f = 0.75$.

6.3 Structural Hardware Security

The structural security metric SHS focuses on the difficulty of understanding a design based on the structural changes induced by logic locking. The more the internal design elements form recognizable and understandable groups that are identifiable by a manual or automated process, the higher the chance for the attacker to uncover the original design's functionality. In this work, we look at two dimensions of structural security: the amount of added change (e.g., number of key gates) and the distribution of the change (e.g., the placement of key gates). These aspects are represented as SHS_{cc} and SHS_{en}.

6.3.1 The Structural Complexity Change

The inherent structural complexity of a design is not sufficient to express a notion of security. Thus, a security measure must include the amount of structural *change* induced by logic locking. We represent this notion with the structural complexity dimension as follows:

Definition 6.7 The **structural complexity change** (SHS_{cc}) represents the structural complexity difference between the original and the logic-locked design.

Before introducing a concrete measure of SHS_{cc}, let us first consider the complexity change in a perfectly secure design in terms of SHS. In this regard, it is desirable to represent a design in a technology-independent format. For example, this can be performed by transforming the design into a multi-level logic network composed of two-input ANDs and inverters, called And-Inverter Graph (AIG). The AIG offers an efficient representation and a good correlation with the expected design size. Within this format, a maximal complexity change, i.e., structural corruption, is achieved if each node (internal gate) in the original graph is covered by at least one key node (key gate). This implies that the output of each gate is affected by at least one key gate. Note that the term *key gate* is a simplified representation of any change that is induced by a logic locking policy. To mirror this notion in terms of structural hardware security, we propose the following:

$$SHS_{cc} = \frac{|\Gamma|}{\sum_{i=0}^{|\mathcal{G}|} FanOut(g_i)}, \tag{6.5}$$

where Γ is the set of key gates and $\mathcal{G}$ the set of original gates. $FanOut(g_i)$ returns the fan-out of a particular gate $g_i \in \mathcal{G}$.

The given definition can result in values larger than 1 if, e.g., all outputs of every g_i are covered with more than one key gate. This would imply an area overhead of more than 100%. However, the introduced dimension should not be limited to a predefined upper bound of an acceptable overhead for security. For now, we assume that any $SHS_{cc} \geq 1$ can be considered as a maximal complexity change.

6.3.1.1 Example: Structural Complexity Change

A simple netlist graph is presented in Fig. 6.4. The total number of key gates is $|\Gamma| = 2$ and the fan-out sum of all gates of the original circuit is $\sum_{i=0}^{|\mathcal{G}|} FanOut(g_i) = 10$. Note that the primary inputs are regarded as gates as well, as they also offer a potential place to insert key gates. The resulting structural complexity change is $SHS_{cc} = 0.2$. It can be observed that the metric incorporates two notions: the amount of added key gates and the amount of affected original gate outputs, thus yielding a value that correlates with the area increase due to the locking mechanism.

Fig. 6.4 Netlist graph
example

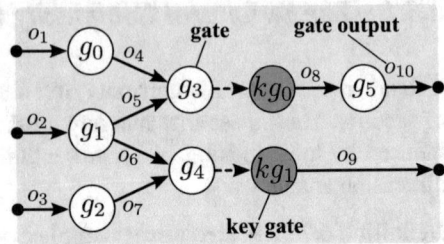

6.3.2   The Structural Key-Gate Entropy

The second dimension of SHS, the structural key-gate entropy metric SHS_{en}, describes how the scheme-induced change, i.e., the set of key gates, is distributed in the design. Intuitively, a uniform distribution of key gates within the original design leads to a greater structural change compared to concentrating key gates only in specific locations.

To enable a generic and adjustable metric, we describe the structural key-gate entropy through a *key-gate distribution measure function* defined as follows:

Definition 6.8 A **key-gate distribution measure function** (f^{kd}) calculates the distribution uniformity of the key gates in a logic-locked design, yielding the result $0 \leq f^{kd} \leq 1$, where a higher value indicates a distribution closer to an optimal uniform distribution.

Thus, the structural key-gate entropy can be defined as $SHS_{en} = f^{kd}$. In principle, this entropy can be measured in a variety of ways.

6.3.2.1   Example: Structural Key-Gate Entropy

For measuring SHS_{en}, we need to be able to quantify the distribution level of the key gates inside the design represented as a Direct Acyclic Graph (DAG). As one implementation example, we introduce an evaluation concept based on topological sorting. The idea of the approach is to extract a representation of the graph that expresses the *spatial distribution* of original and key gates within the netlist. We refer to the extracted spatial information as the *structural distribution vector*. Let us consider the examples shown in Fig. 6.5. Here, the netlist is presented in the form of a DAG with the actual key inputs omitted. After the topological sort is performed, the distribution vector is extracted by measuring the distances between the location of each key gate in the topological sort. For example, in Fig. 6.5a, the gates g_6, g_7, and g_9 are marked as key gates. Thus, the extracted distribution vector is $s_d =$ [6, 0, 1, 0], where each element represents the distance to a *neighboring* key gate. The extracted vector clearly indicates a suboptimal distribution, as a large cluster of original nodes (g_0 to g_5) is present. Thus, a large part of the netlist has not been structurally affected by the locking scheme. After the extraction is done, the next

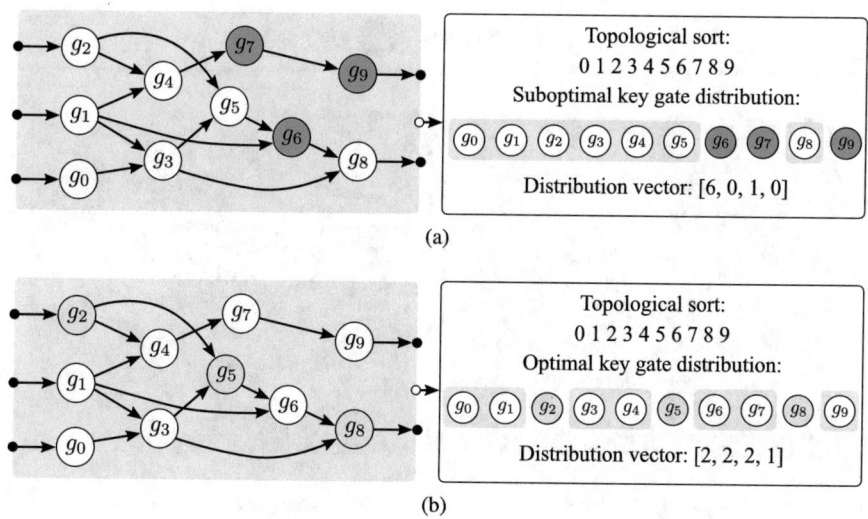

Fig. 6.5 Topological sort-based structural key-gate entropy calculation: (**a**) suboptimal and (**b**) optimal solution

step calculates the distance to an optimal distribution of key gates, s_{opt}. An optimal solution contains evenly distributed key gates throughout the netlist structure. In this case, $s_{opt} = [2, 2, 2, 1]$. The final SHS_{en} is the d_e between the extracted and the optimal vector, scaled to the maximum possible distance:

$$SHS_{en} = \frac{d_e(s_{opt}, s_{max}) - d_e(s_{opt}, s_d)}{d_e(s_{opt}, s_{max})} = 0.195,$$

where d_e is the Euclidean distance of two vectors. In this example, one maximal distance vector is $s_{max} = [7, 0, 0, 0]$, which corresponds to all key gates being focused on the primary outputs.

A better solution is presented in Fig. 6.5b. Here, the key gates are represented by the nodes g_2, g_5, and g_8. This configuration results in an optimal distribution ($SHS_{en} = 1$). The uniformity comes from the presence of smaller gate clusters of approximately the same size. Therefore, the available key gates are distributed to induce a maximal structural change to the original design. As SHS_{en} is scaled according to s_{max}, it always yields a value from 0 to 1.

6.3.3 The Problem of Multidimensionality

Interestingly, there is no explicit relation between SHS_{cc} and SHS_{en}. For example, if $SHS_{cc} = 1$, approximately half of all the gates of the design are key gates

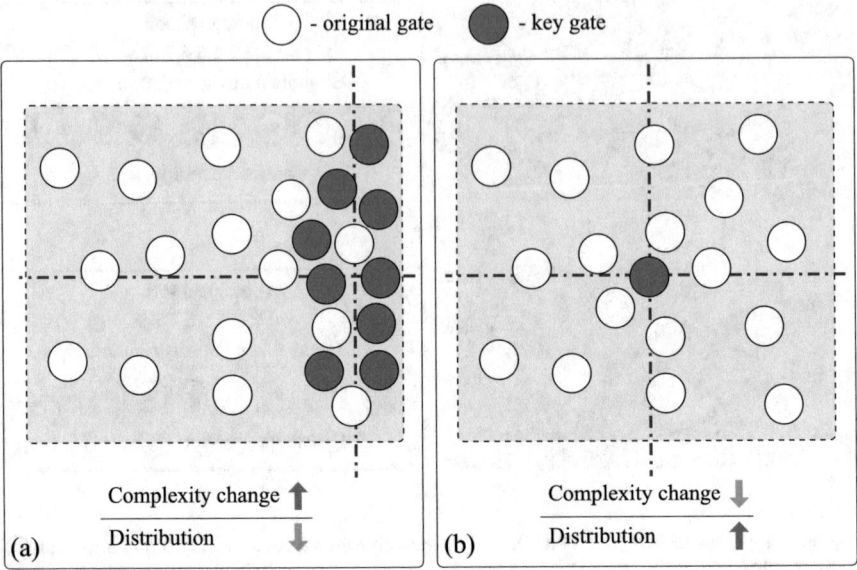

Fig. 6.6 Example of the multidimensionality problem: change vs. distribution

(assuming the measuring defined in Eq. 6.5). This does not imply that the key gates are distributed uniformly. In fact, all key gates might be spatially concentrated at the primary inputs or outputs without significantly corrupting the original design, as visualized in Fig. 6.6a. On the contrary, $SHS_{en} = 1$ can be achieved by placing a single key gate in the topological center of the original design, as shown in Fig. 6.6b. Naturally, a single key gate results in an insignificant change in complexity. Therefore, it is important to measure both dimensions in order to fully explore structural hardware security. In addition, a correlation between functional and structural security dimensions must not necessarily exist. For example, high structural security does not imply security against attacks that exploit the functional dimension of a design and vice versa. As discussed at the beginning of this chapter, this is one of the main reasons why hardware security metrics for logic locking are undoubtedly a challenging problem.

6.3.4   Emerging Dimensions

The dimensions visualized in Fig. 6.1 are expected to evolve with emerging attack vectors. For example, novel Machine Learning (ML)-based attacks in the oracle-less model are capable of predicting a correct key through the analysis of the added key-dependent change. Thus, these attacks exploit a structural dimension which is not covered neither by the complexity change nor by the key-gate entropy. Therefore,

Table 6.1 ISCAS'85 benchmarks used for evaluation

IC	#inputs	#gates	#outputs	IC	#inputs	#gates	#outputs
c499	41	202	32	c2670	233	1193	140
c432	36	360	7	c3540	50	1669	22
c880	60	401	26	c5315	178	2307	123
c1355	41	546	32	c7552	207	3512	108
c1908	33	880	25				

the security dimensions of logic locking will have to be continuously adjusted and expanded to accommodate for new attack targets.

6.4 Evaluation

To showcase the application of the proposed metric dimensions, we present an evaluation for a set of prominent schemes from the pre- and post-SAT era.

6.4.1 Experimental Environment

We implemented the dimension evaluation methodologies in the form of a software library based on Python and the open-source Icarus Verilog [199] simulation tool. The online-available attack implementations provided by Subramanyan et al. [175] were utilized for the evaluation of f^{dc} functions, including the original and partial-break SAT attack, the PSA, and the hybrid PSA/partial-break attack. The evaluation timeout for the SAT attack was set to 10 h.[3] The total number of input key pairs for the evaluation of D^f, C^f, and S^f was set to 10^6. The evaluation was performed on the benchmarks presented in Table 6.1. All experiments were run on an Intel i5-4460 CPU with 8 GB of RAM.

An average security analysis across the *entire* benchmark portfolio has been performed for two cases: evaluation of pre-SAT schemes and a comparison of pre- and post-SAT schemes. All schemes were deployed by limiting the area overhead to 25%.

---

[3] This time limit is suggested in [175]. The assumption for this limit is that the probability is high that the attack is collapsing to a brute-force search if the attack fails within this time frame. In most cases, vulnerable schemes can be broken within a few minutes. Note that the time limit is also bound by the time the locked design spends in the hands of untrusted parties.

6.4.2   Results: Pre-SAT Comparison

The following pre-SAT schemes have been selected for evaluation: Strong (Secure) Logic Locking (SLL) [129], AND–OR locking [53], and FLL using XOR gates [132]. More details can be found in Sect. 5.2.

The average evaluation results are shown in Fig. 6.7a. Even though all schemes achieve only a very low KSS value,[4] some differences can be identified. For example, the XOR/XNOR placement in SLL offers more security than the XOR insertion in FLL-XOR. Furthermore, both SLL and FLL-XOR are more secure than AND–OR locking. This conforms with the following. First, the SLL scheme aims at achieving a complex interference among key gates, whereas FLL-XOR selects insertion locations with the highest fault impact, thus focusing on output corruptibility. Second, the insertion of AND, OR, and MUX gates generally leads to clauses that are easier to satisfy with the SAT attack compared to XOR/XNOR gates [171, 175]. Consequently, as the most efficient f^{dc} is the SAT attack, as expected, SLL performs best in terms of KSS.

Moreover, the highest deceptiveness D^f is reached for SLL. This is, in fact, an expected result as SLL performs best against the SAT attack. Thus, the deceptiveness correlates positively with SAT resilience. For AND–OR and FLL-XOR, the deceptiveness is close to zero, confirming the low KSS values. The same conclusions can be made through the corruptibility dimension.

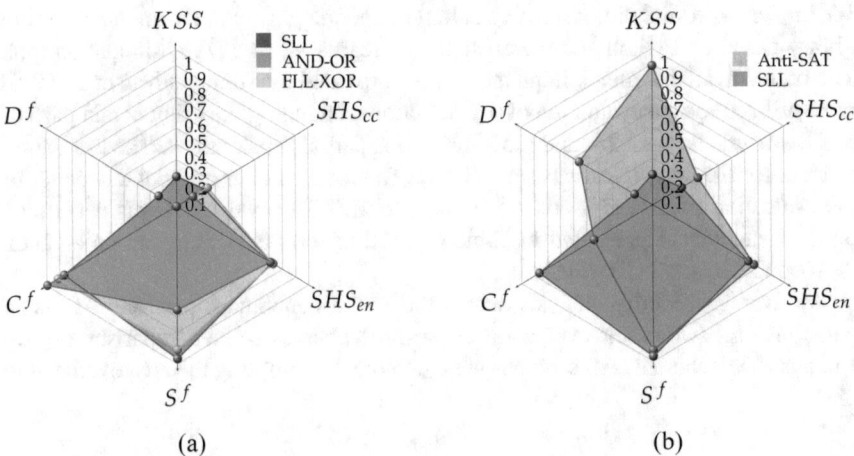

Fig. 6.7 Security metrics evaluation: (**a**) pre-SAT average comparison and (**b**) pre/post-SAT average comparison

---

[4] Reminder: KSS is gradually decreased by the application of all attacks that lower the key-space size.

The secrecy S^f for SLL and FLL-XOR is, on average, higher compared to AND–OR locking. This result can be traced back to the objectives of the schemes. SLL creates a strong interference between key gates. FLL-XOR tries to achieve high output corruption. On the contrary, AND–OR locking minimizes rare values in a netlist. Thus, incorrect keys for SLL and FLL-XOR are more likely to generate more corrupted and well-distributed outputs.

The complexity change SHS_{cc} is around 0.2 for all schemes, confirming its relation to the selected overhead of 25%. The key-gate entropy is around 0.75 on average, confirming the relatively high spatial distribution of key gates within all schemes.

6.4.3   Results: Pre/Post-SAT Comparison

To compare pre- and post-SAT schemes, we present the evaluation of Anti-SAT [206] and SLL. Anti-SAT belongs to the post-SAT era and exhibits a high functional resilience against the SAT attack. The evaluation results are shown in Fig. 6.7b.

The average key-gate entropy is slightly higher for Anti-SAT compared to SLL. At first glance, this is misleading since Anti-SAT is typically implemented by inserting an additional functional block into the design. However, as this can easily lead to removal attacks [216, 217], Anti-SAT is often combined with Random Logic Locking (RLL). Therefore, in this case, a higher SHS_{en} is expected.

The combination with RLL has, in theory, a negative impact on the deceptiveness factor. However, to exemplify the deceptiveness nature of Anti-SAT, the evaluation of D^f masks the effect of RLL-induced key gates. Thus, the relatively high value of $D^f = 0.55$ is representative and expected for the SAT-resilient scheme. Note that D^f is likely to be even higher in case Anti-SAT is not masked by RLL. Nevertheless, the achieved D^f represents a lower bound.

The SAT resilience of Anti-SAT is clearly visible in the high value of $KSS = 0.89$, outperforming SLL ($KSS = 0.19$) by more than 368%. The average Anti-SAT complexity change of $SHS_{cc} = 0.34$ is 61.9% higher compared to SLL, indicating the additional price for SAT resilience.

For both schemes, the secrecy S^f seems to be high enough to ensure security. However, since heuristic attacks based on secrecy have not been introduced yet, further evaluations will have to be performed to validate the security level.

6.5   The Security-Cost Trade-Off Problem

An important aspect in designing secure HW is the security-cost trade-off. Thus far, a consensus on the acceptable logic locking overhead has not been reached. Typically, existing proposals assume overhead limits below 50%. Thus, the security of locking schemes is often evaluated only within this cost frame. However, limiting the cost overhead can lead to a skewed security evaluation. This is further exacerbated by the use of relatively small benchmarks. In this context, a low area overhead leads to very little change within small designs, thus not allowing a scheme to exercise security properties for a higher cost.

6.5.1   Case Study: Overhead Implication on Security

To gain more insights into the problem of the security-cost trade-off, we present a case study based on the RLL scheme and the prominent SAT attack under an oracle-guided attack model. Hereby, the security evaluation focuses exclusively on the key retrieval, i.e., on the impact of the attack on the key-space size for high area overheads. The RLL scheme has been chosen due to the following. First, the core of the scheme is based on XOR/XNOR key-gate insertion. As this is a fundamental component of a large variety of schemes, RLL is a suitable and representative choice. Second, XOR/XNOR-based schemes are typically vulnerable to the SAT attack. Thus, evaluating the impact of a higher cost budget on the security of the RLL scheme can provide compelling insights.

6.5.1.1   Experimental Environment

A subset of the ISCAS'85 [31] and MCNC [210] benchmark was chosen for evaluation. All benchmarks were locked with RLL for 50%, 60%, 70%, 80%, 90%, and 100% area overhead. The attack timeout was set to 10 h (see Sect. 6.4.1). All presented results are sorted according to the netlist size.

6.5.1.2   Evaluation Results

The results of the attack-resilience evaluation are shown in Fig. 6.8. The resilience rate indicates the total percentage of *unbroken* locked circuits per circuit size categories. The results indicate a clear trend; across all designs and size categories, the number of unbroken circuits steadily increases (up to 80% for designs with more than 3000 gates). Similar observations can be made for other oracle-guided attacks as well. More details are provided in Appendix D.1.

6.5.2   Discussion

The results of the case study clearly indicate that a higher cost budget can significantly impact the security evaluation for a specific scheme. Thus, limiting

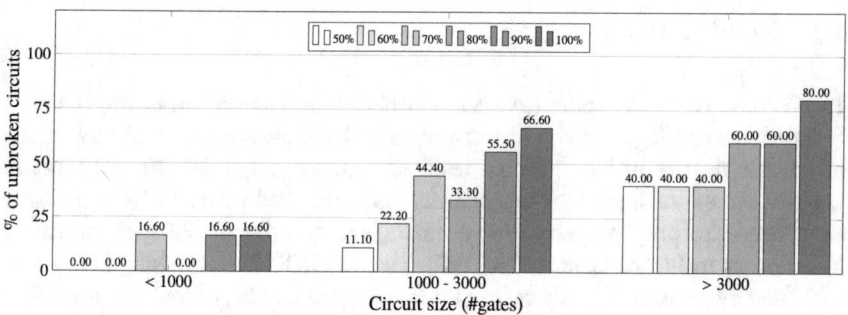

Fig. 6.8 Resilience against the SAT attack per circuit size categories

the evaluation to only a few percent in acceptable overhead can lead to distorted security assessments. Furthermore, common evaluation benchmarks are often far from being representative, as modern HW designs are typically much more complex. Thus, a low cost budget and low-complexity benchmarks greatly limit the frame of the security evaluation.

Moreover, from a research point of view, the security evaluation should never be limited by a low cost budget. In reality, an acceptable overhead is largely driven by industry, i.e., the needs of customers and specific use cases of the target HW. Specifically, in security-critical applications, where logic locking is most likely to be deployed, hardware redundancy is a common practice and security objectives outweigh the increased cost.

6.6 Limitations and Outlook

The presented security dimensions aim at capturing all necessary security aspects of logic locking. Even though novel attacks and protection policies are regularly being introduced and revised, thus far, all attacks have been exploring either a structural or a functional aspect of logic locking. Therefore, the presented security dimensions offer a comprehensive foundation for future metric designs. Nevertheless, future adaptations will undoubtedly be required to accurately describe and assess security-relevant features of contemporary locking schemes. In this regard, the security metric system is easily adaptable by either adjusting the way a specific dimension is computed or by extending the model with novel constructs. In any case, the future of hardware security metrics is, most likely, multidimensional in nature.

6.7 Related Work

To this day, very few works have addressed the challenge of measuring security
in the context of logic locking. Most proposals have measured security by means
of specific metrics or based on the resilience against only one type of attack. In
the pre-SAT era of logic locking, security was measured by exclusively focusing
on a single security property. For example, the security of SLL was measured
based on its resilience against PSA [129]. The AND/OR-based locking policy was
evaluated by measuring its success in minimizing the number of low-controllability
signals [53]. The security of FLL was determined by the Hamming distance between
the correct and incorrect output for incorrect activation keys [132]. In the post-SAT
era, measuring the security of logic locking focused on the resilience against the
SAT attack. For example, a SAT-attack-based theory on logic locking for capturing
the design space and trade-off between the error rate and attack complexity was
presented in [227]. Recently, a more elaborate metric was introduced in [75]. Here,
the security of logic locking is evaluated by measuring three properties: functional
corruptibility, approximate and exact SAT-attack resilience, and removal resilience.

In comparison, the metric system discussed in this book incorporates both
functional and structural aspects of locking policies, thereby allowing for the
inclusion of future extensions and more precise evaluation procedures.

6.8 Synopsis

This chapter introduced one of the first hardware security metrics for logic locking.
The metric is designed to capture structural and functional security features in a
multidimensional model. Herewith, each dimension is extensible and adjustable to
account for the rapidly changing landscape of logic locking design. Furthermore,
a case study on the impact of the cost budget on the security assessment of logic
locking has been introduced, showcasing the importance and necessity of allowing
higher overheads in a comprehensive security evaluation. In the following chapters,
where necessary, only parts of the metric system are required, as further discussions
focus on the specifics of the assumed attack scenario.

Part III
Logic Locking in Practice

Chapter 7
Software Framework

As elaborated in Chap. 5, the past two decades of research have brought forth a vast landscape of logic locking schemes. Consequently, novel key recovery attacks have been subsequently introduced, creating a repeating design–attack–improve cycle. Even though this process has uncovered many critical security aspects of modern locking schemes, the existing proposals have long been designed without keeping the intrinsic features of the underlying Hardware (HW) in mind. This has led to inflexible and overspecialized logic locking, thus creating a growing gap between the technology and its applicability to silicon-ready designs.

Modern HW typically consists of multiple, functionally and structurally isolable components. This book refers to these components as *modules*. As an example, in a processor design, a module can be the instruction decoder, flush controller, Arithmetic Logic Unit (ALU), and others. The existing locking proposals have often neglected this aspect of hardware, thereby treating the complete HW design as a single unit, e.g., in the form of a flattened gate-level netlist. This disables the influence of expert knowledge about the HW during the deployment of logic locking. For example, some HW components might be more vulnerable to specific attack vectors, requiring dedicated security policies. Moreover, by adjusting the logic locking overhead in terms of power, area, and delay based on the modular nature of the HW design, the incurred cost can be steered to accommodate the desired requirements. If the inherent structural and functional components of the initial HW design are ignored, a targeted and application-specific protection scheme becomes significantly more difficult to implement. Moreover, the process of deploying logic locking must be compatible with the subsequent HW design and fabrication stages to allow a seamless integration in a common industrial setting. A locked design must be able to continue in both the Application-Specific Integrated Circuit (ASIC) and Field-Programmable Gate Array (FPGA) flow without imposing specific technology requirements.

To enable a practical deployment of logic locking, in this chapter, we intro-duce the design and implementation of a modular, industry-proven, technology-

© The Author(s), under exclusive license to Springer Nature Switzerland AG 2023
D. Sisejkovic, R. Leupers, *Logic Locking*,
https://doi.org/10.1007/978-3-031-19123-7_7

independent logic locking framework for the protection of multi-module HW designs. Herein, the framework enables a modular deployment of locking policies depending on the HW features and the application scenario.

This chapter is organized as follows. An overview of the framework is presented in Sect. 7.1, while Sects. 7.2–7.6 elaborate on each stage of the framework in more detail. Section 7.7 discusses the limitations and outlook. Finally, a summary of this chapter is given in Sect. 7.8.

7.1 Framework Overview

An overview of the proposed logic locking framework is shown in Fig. 7.1. The input to the framework is a Register-Transfer Level (RTL) design instance. The output consists of an activation key for the locking mechanism and a set of locked technology-independent gate-level netlists that are integrated into the final protected design.

The framework consists of five major phases [167]: (1) The module selection phase (Fig. 7.1a) in which trusted personnel selects which components of the HW design should be included for protection. (2) The module preprocessing phase (Fig. 7.1b) prepares the selected modules for logic locking via a transformation to a generic, technology-independent gate-level format. (3) The logic locking phase (Fig. 7.1c) loads the processed netlists and deploys the selected locking policy. The output of this phase is the activation key and a set of locked netlists. (4) The netlist integration phase (Fig. 7.1d) accounts for two steps. First, the selected components

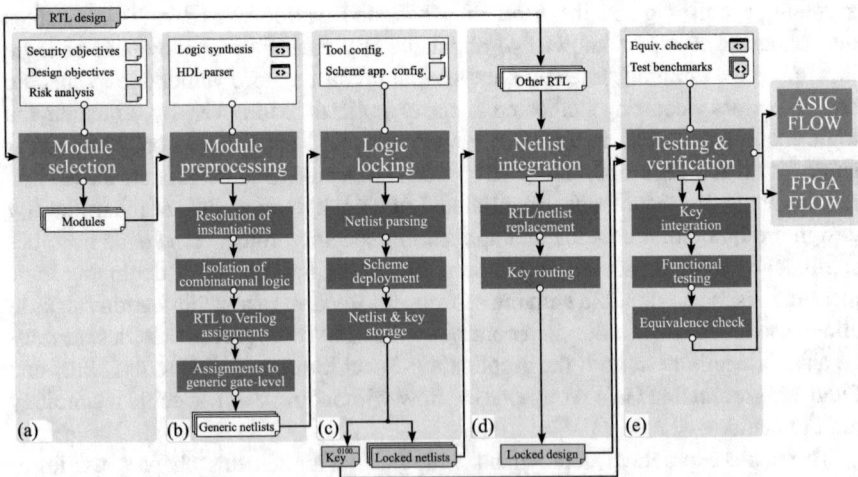

Fig. 7.1 The logic locking framework: (**a**) module selection, (**b**) module preprocessing, (**c**) logic locking, (**d**) netlist integration, and (**e**) testing and verification

from the first phase are replaced by the locked netlists. Hereby, other unprotected RTL modules of the HW design are given as input. Second, the key routing is performed to create the necessary wiring between the primary key inputs and the locked components. (5) The last phase of the flow is in charge of the testing and verification procedure (Fig. 7.1e). Finally, due to the technology independence of the framework's output format, the locked design can proceed in both the ASIC and FPGA flows.

The following sections elaborate on all five phases in more detail, thereby referencing the flow in Fig. 7.1. Herewith, the necessary inputs, tools, and requirements are listed in each section separately.

7.2 Module Selection

The module selection phase incorporates the selection of suitable HW design modules that are intended for inclusion in the logic locking procedure (Fig. 7.1a). The input to this phase is the complete RTL design of the target HW. So far, the selection process has to be performed manually by trusted in-house personnel. In terms of the Verilog hardware design language, a module defines an enclosed (System) Verilog file. This definition already sets a certain granularity for module separation that is defined by the implementation manner of the specific HW design. As previously mentioned, processors typically consist of components that are commonly known across different microarchitectural implementations, such as decoders, ALUs, register files, and others. Whether a component is seen as a module at this level or at another granularity (e.g., at pipeline stage level) is open to interpretation.

The main challenge in the module selection is to decide which components are relevant for protection. Unfortunately, as usual in security-related issues, there is no exact solution. However, performing a risk analysis is a common tool in security-driven system design. The same approach can be applied to the selection of relevant design modules. The risk analysis can potentially prioritize modules based on their relevance in enabling certain attacks. For example, a System-on-Chip (SoC) typically consists of a Central Processing Unit (CPU), memory, peripherals, and other components. Depending on the outcome of the risk analysis, it might be relevant for a particular HW design to only protect a subset of the overall system. As discussed in Sect. 2.1.2, in the case of an intelligent hardware Trojan embedded in an SoC, it might be relevant to only protect the CPU, in case the risk analysis concludes that infecting other components would not lead to a relevant form of attack. The analysis is driven by two, often opposing objectives:

1. **Security objectives:** The decision on whether a module must be protected or not can be influenced by the specific security objectives driven by the attack model. For example, as discussed in Sect. 5.1, specific attacks, such as the Path-Sensitization Attack (PSA), have to be able to control and observe specific

combinational paths in the activated Integrated Circuit (IC) to retrieve the key value. If this attack is considered in the attack model, it is crucial to enable the protection of all modules that are easily accessible via the primary I/Os or through the scan chain. Moreover, the module selection must also be driven by knowledge of the application environment in which the HW design will be placed after production. For example, the deployment environment can define what kind of access an attacker has to the activated IC. This can include physical in-the-field access or remote access (via a specific communication protocol). Among others, these aspects can impact the decision on what HW components need to be protected.

2. **Design objectives:** Another important role is played by the design objectives. Typically, the HW design has to fulfill certain area, power, and performance requirements for a predefined use case, all of which have an impact on the final cost of the produced IC. As the logic locking mechanism induces a certain cost overhead, the HW design team has to perform a security–cost trade-off analysis. This can lead to adjusting the amount of locking-induced change in the HW to achieve the desired objectives.

Once identified, the selected modules proceed to the module preprocessing phase of the framework. Other RTL modules that have not been selected for locking remain unaffected by the tool flow of the framework. These components are included only in the netlist integration phase to assemble the final locked design.

7.3 Module Preprocessing

The module preprocessing phase prepares the selected modules for the locking procedure (Fig. 7.1b). The input consists of the selected modules in RTL format. The preprocessing transforms the input modules into generic gate-level netlists. The term *generic* refers to the technology-independent format of the output files. In this book, a generic netlist is represented in the form of primitive gate types defined in the Verilog Hardware Description Language (HDL) standard. Note that only a generic format ensures that the final locked design can proceed in both the ASIC and FPGA flows. Moreover, the locked netlists can easily be processed with any Verilog-compatible open-source or commercial simulation, synthesis, and verification tools without creating a dependency on specific technology libraries.

The preprocessing phase consists of four steps: (1) First, inter-module instantiations have to be resolved. This step gives control over which part of a module hierarchy should be included in the locking mechanism. (2) The next step includes the isolation of combinational logic into separate modules. (3) Afterward, the prepared input RTL is transformed into generic Verilog-only assignments with the use of a logic-synthesis tool. (4) Finally, the Verilog assignments are transferred into a generic gate-level format. Note that the concrete implementation of the four steps can be adjusted to the underlying logic-synthesis tool. This book makes use of the Synopsys Design Compiler (DC) [179] as it is well adopted by industry due to its proven capabilities to handle complex HW designs. However, the general flow can be transferred to other logic-synthesis tools, such as the Yosys open synthesis suite [200]. The following gives more details on each step.

7.3.1   Resolution of Instantiations

RTL implementations are typically written in a modular and hierarchical fashion. Herewith, one module can include multiple instantiations of other modules within its body. For example, the execution stage of a processor pipeline can include the instantiation of an ALU module. Depending on the implementation details, the ALU itself can embed separate units for addition, shifting, comparison, floating-point operations, and others. This creates a hierarchical design structure that can be adjusted to the desired granularity for the locking procedure. As an example, let us consider two cases. In the first case, the risk analysis has concluded that it is only relevant to lock the ALU without impacting the floating-point unit. Here, the unit can be temporarily removed (or commented in code) to avoid being translated in the next steps. After locking is done, the missing module can be re-embedded into the HDL code. The second case includes fully locking the entire ALU with all its sub-modules. This implies that the internal module instantiation must be resolved. This can be performed through netlist *flattening* by using the desired logic-synthesis tool. The flatting process merges all structural hierarchies in the HW design into a single netlist.

7.3.2   Isolation of Combinational Logic

As discussed earlier, a generic netlist format offers many advantages. Through the use of DC, an RTL design can be translated into Verilog-only code. However, this is not the case for sequential elements as their implementation remains in the form of instantiations of technology-dependent registers. Consequently, if the sequential elements remain in the netlist, the final modules create a reliance on a specific technology. In an ASIC-only flow, this reliance creates no limitation as the design is always recompiled for a specific technology node. However, emulating the HW design on an FPGA board is not possible as a technology dependence exists. Therefore, the process of isolating combinational logic can be skipped depending on the desired flow.

A simple example of the isolation of combinational logic is presented in Fig. 7.2. Here, the combinational logic of the original flush controller (Fig. 7.2a) of the Ariane RISC-V core [225] is isolated by moving the logic into the separate module `controller_logic`. To ensure the same functional behavior, the isolated logic must be instantiated in the original controller.

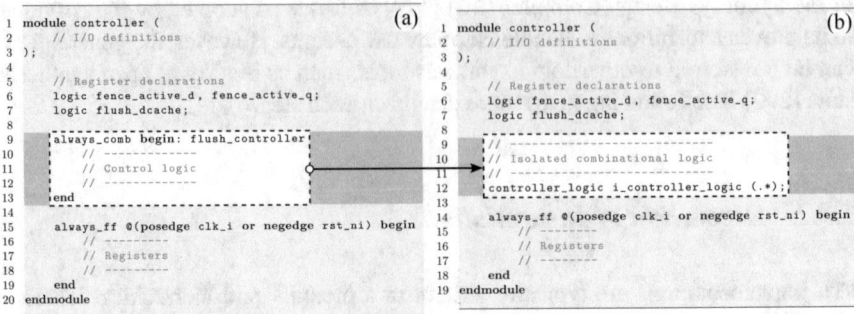

Fig. 7.2 Example of isolation of combinational logic: (**a**) initial module and (**b**) isolated logic

```
 1  module m (                    (a)
 2      input A,
 3      input B,
 4      input C,
 5      output reg F
 6  );
 7
 8      always @ (A or B or C) begin
 9          if (A == 0)
10              F = B;
11          else
12              F = C;
13      end
14  endmodule
```

```
 1  module m ( A, B, C, F );        (b)
 2      input A, B, C;
 3      output F;
 4
 5      assign F = ((-(A) & B) | (A & C));
 6  endmodule
```

Fig. 7.3 Example of RTL to Verilog assignments mapping: (**a**) initial RTL and (**b**) Verilog assignments

7.3.3 RTL to Verilog Assignments

After the combinational path has been isolated where necessary, the selected modules are further processed with DC. This synthesis tool offers the option to map the logic to the generic technology library GTECH, meaning that the included gate types have no concrete timing, power, or delay specifications. If a concrete technology library is selected beforehand, it can be already integrated at this point in the framework. Once DC maps the loaded modules to GTECH, the tool can be instructed to translate the HW design in the form of Verilog-only assignments. As discussed in Sect. 7.3.2, this is only applicable for combinational logic. The translation process is rather simple. The used synthesis library contains the Boolean equation for each gate type, i.e., logic cell. By simply traversing the netlist graph, the Boolean function of each cell is extracted and stored into the final Verilog file. This process yields a Verilog module written in the form of simple assignments. An example of the input and output of this procedure is shown in Fig. 7.3.

7.3.4 Assignments to Generic Gate Level

The final step of the module preprocessing phase is the transformation of assignment-level Verilog into a gate-level netlist. The input to this step consists of the DC-generated Verilog modules. Here, the core idea is to represent the Verilog assignments in the form of primitive gate types (AND, OR, XOR, NOT, and BUF), thereby creating a representation in the form of a directed graph. As discussed in Sect. 4, locking schemes typically operate on a gate-level netlist through the insertion, replacement, or transformation of gates, i.e., graph nodes. Therefore, having an easy-to-process graph representation of the HW design matches the working principles of the logic locking procedure.

The transformation procedure is exemplified in Fig. 7.4. It consists of two steps: (1) First, the input Verilog RTL code is parsed, and a lexical and syntax analysis is performed (Fig. 7.4a). This step is based on the open-source library Pyverilog [182]. The output of this step is a file containing the Abstract Syntax Tree (AST) of the input RTL in simple textual format. This format allows for simple parsing of the AST components in the next step. (2) Next, the netlist generation is performed (Fig. 7.4b). Here, the input AST file is loaded into a tree-based memory representation, where each *internal* node represents a primitive gate type and all *leaf* nodes reference primary inputs or internal wires (outputs of internal nodes). Finally, the in-memory tree is traversed, and code is emitted for each node. Herewith, one internal node generates one gate definition in the final netlist. The netlist generation has been implemented in a standalone C++ based application that is able to directly invoke the Python-based code parser. Note that the AST input file contains all necessary information to reconstruct all module components. Therefore, the created netlist is a functionally valid Verilog module, equivalent to the input RTL.

The discussed translation process has a profound impact on the applicability of logic locking in later stages. As the input RTL is mapped to a generic primitive gate format without further optimizations, the netlist embodies the HW design at the

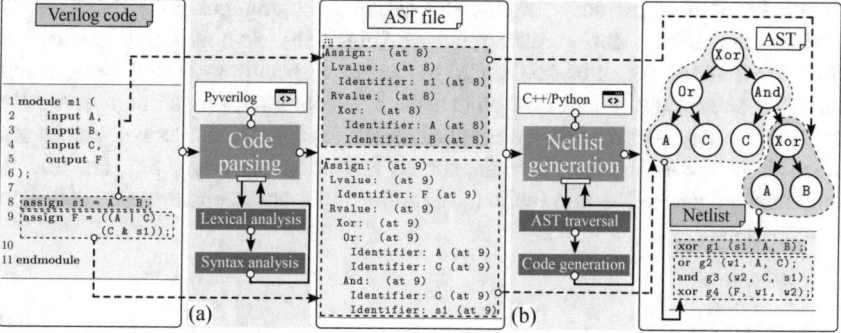

Fig. 7.4 Example of Verilog assignments to generic gate-level netlist transformation: (**a**) code parsing and (**b**) netlist generation

lowest possible granularity. In case a specific technology library is used alongside the optimization, the output netlist would entail design components at a higher level, i.e., specific cells that encapsulate part of the HW design's functionality. A low granularity maximizes the opportunity for logic locking to induce functional and structural changes. One enabling factor for the higher granularity is the way the in-memory AST is assembled; each primitive gate has at most two inputs (except NOT and BUF). This format maximizes the spread of functionality by not focusing gate trees into a single gate, e.g., an N-input AND gate is represented as a tree of 2-input ANDs. Once the RTL assignments have been translated to generic gate-level netlists, the generated files are ready for the logic locking stage of the framework.

7.3.4.1   Translation with a Limited Technology Library

The introduced process is not the only option of translating a gate-level netlist into a generic 2-input gate format. Another possibility is to utilize a technology library by limiting the synthesis tool to use only 2-input gates. Thus, the synthesis process is steered toward mapping the design to the desired format. Afterward, a simple post-processing step can map the technology-dependent cell instantiations to primitive Verilog gates. However, this procedure limits the optimization opportunities during logic synthesis.

7.4   Application of Logic Locking

The central part of the framework lies within the application of logic locking (Fig. 7.1c). The input to this stage is the set of selected and preprocessed netlists as discussed in Sects. 7.2 and 7.3. The logic locking stage consists of three steps as visualized in Fig. 7.5: (1) First, the netlist parsing takes care of loading the input netlists into memory (Fig. 7.5a). (2) After the netlists have been represented in the form of in-memory graphs, the selected locking policy is deployed to perform the desired design manipulations (Fig. 7.5b). At this point in the tool flow, the framework can be configured to execute a specific locking scheme based on the desired security and cost objectives. The configuration can be specifically adjusted for any subset of the selected modules. (3) Finally, the locked graphs are stored in netlist format alongside the correct activation key (Fig. 7.5c). The flow is implemented using C++ and Python. The following sections elaborate on each step in more detail.

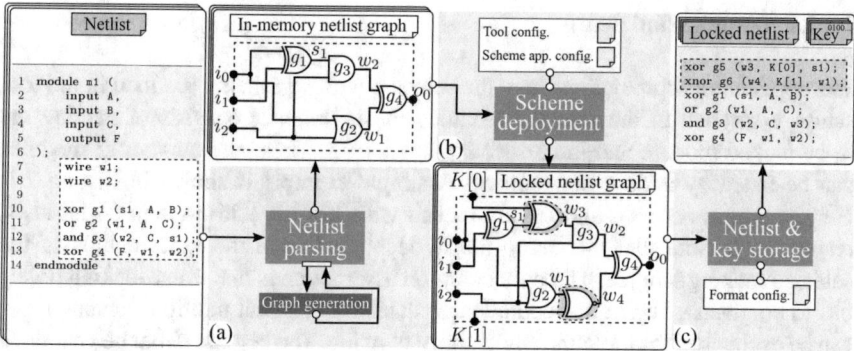

Fig. 7.5 Application of logic locking: (**a**) netlist parsing, (**b**) scheme deployment, and (**c**) netlist and key storage

7.4.1 Netlist Parsing

Before a selected logic locking scheme can be deployed, the input netlist files have to be parsed and loaded. As shown in Fig. 7.5a, the netlists are represented as graphs in memory, where the gates, registers, and primary Inputs and outputs (IOs) are modeled as node *objects*. Each node has a reference to its predecessor and successor nodes. This netlist representation enables the implementation of schemes as graph manipulation algorithms. Designing the graph using Object-Oriented Programming (OOP) concepts makes the analysis and manipulation process of logic locking schemes easy to implement, adapt, and maintain.

7.4.2 Scheme Deployment

After the netlists have been loaded, the selected locking scheme is deployed to perform HW design manipulations according to the provided configuration files (Fig. 7.5b). The configuration consists of the tool and the application setup file. Both configuration files are provided as command arguments to the framework tool.

7.4.2.1 Tool Setup

The tool setup concerns the overall configuration parameters of the entire tool. It consists of two sections. The first section concerns general setup information, such as wire and node naming rules, input and output paths, paths to external libraries, and others. The second section controls the setup details of logic locking, including the scheme type and its specific adjustments. All supported configuration entries can be found in Appendix B.1.

7.4.2.2 Application Setup

The application setup defines how the selected locking scheme (defined in the tool setup) is applied to the selected modules. A single entry consists of naming the to-be-locked module and the respective key length. Thus, any number of modules can be listed by adding new entries. A simple example is shown in Listing 7.1. Here, modules m_1, m_2, and m_3 are locked with 128-bit, 256-bit, and 64-bit keys, respectively. Note that this setup file is decoupled from the exact details of the selected locking scheme. The only common assumption is that every locking policy has to be given a key of a specific length. This implies that multiple scheme types can be exchanged for the same application setup file. The key can either be generated automatically based on a random source or manually provided as an input parameter.

Listing 7.1 Application setup file.

```
1   # this is a comment
2   # <source_module> <key_length>
3   m1 128
4   m2 256
5   m3 64
```

7.4.2.3 Design Concept

One important aspect of a holistic framework is its modular implementation. As new security vulnerabilities are uncovered, the existing locking policies are being revised and new ones are introduced. Therefore, the presented framework is designed to offer a simple programming interface based on OOP concepts for easier integration of novel schemes. The design concept is built on modeling each scheme component (e.g., locking algorithm, node selection, node insertion, key generation, and others) as a separate class. This modular implementation allows for an ecosystem for fast-tracked scheme prototyping. Using the discussed framework, throughout the course of this book, multiple logic locking schemes have been implemented and evaluated against relevant attack vectors. The adaptations can be controlled via the tool setup file. More details on the design concept are available in Appendix B.2.

7.4.3 Netlist and Key Storage

The final step in the logic locking stage is the storage of the generated outputs, as visualized in Fig. 7.5c. The output consists of the locked gate-level netlists and a set of activation keys in binary string format. The resulting netlists must fully adhere to the Verilog standard. Therefore, the scheme deployment ensures that every inserted key-driven structure is properly embedded in the netlist graph. The key input for each netlist is defined as a Verilog array in the form of input

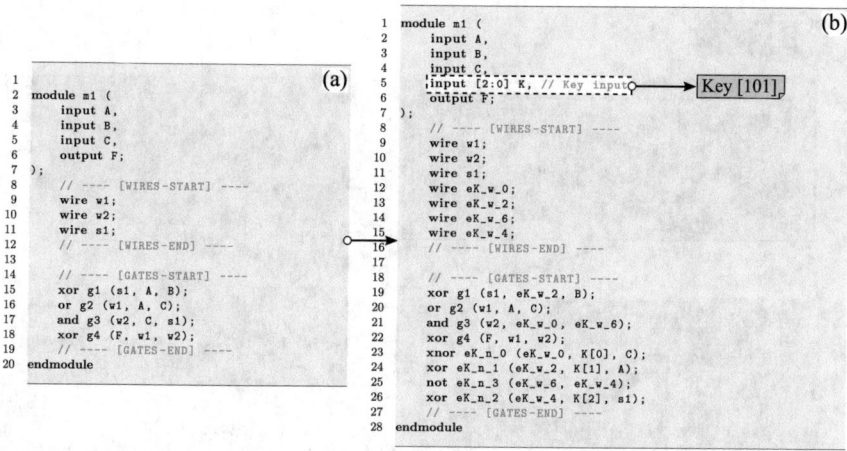

Fig. 7.6 Example of the locked outputs: (**a**) initial and (**b**) locked gate-level netlist

[N-1:0] <key_name> for a key length of N. This input is registered as a primary input of the module. A simple example of a netlist locked with Random Logic Locking (RLL) using a 3-bit key is presented in Fig. 7.6. Note that the output netlists can, in principle, be stored in any necessary format based on the in-memory graph representation of the HW design. Currently, the framework supports the .v and .bench format. Due to its simplicity, the latter is often used in open-source Electronic Design Automation (EDA) tools. With the storage of the locked netlists and the activation keys, the locking procedure continues to the integration stage, as discussed in the following.

7.5 Integration

The next stage of the framework concerns the integration of all locked modules into the initial HW design (Fig. 7.1d). As mentioned in Sect. 7.2, not all modules need to be involved in the logic locking process based on the initial risk analysis. Consequently, the final locked design is composed of locked netlists as well as other unchanged RTL modules. The integration consists of two steps: (1) the RTL/netlist replacement and (2) key routing.

The first step includes a simple replacement of all initial RTL modules with their locked netlist counterparts. Afterward, the second step ensures the inclusion of correct in-code key routes between the primary key input and the locked modules. Since a locked netlist can be at an arbitrary level in the design hierarchy, the key array must be routed through multiple levels. A simple visualization of this process is shown in Fig. 7.7. Here, the complete key provided to the top-module Top M is partitioned and forwarded to the destination modules. If a locked module is at

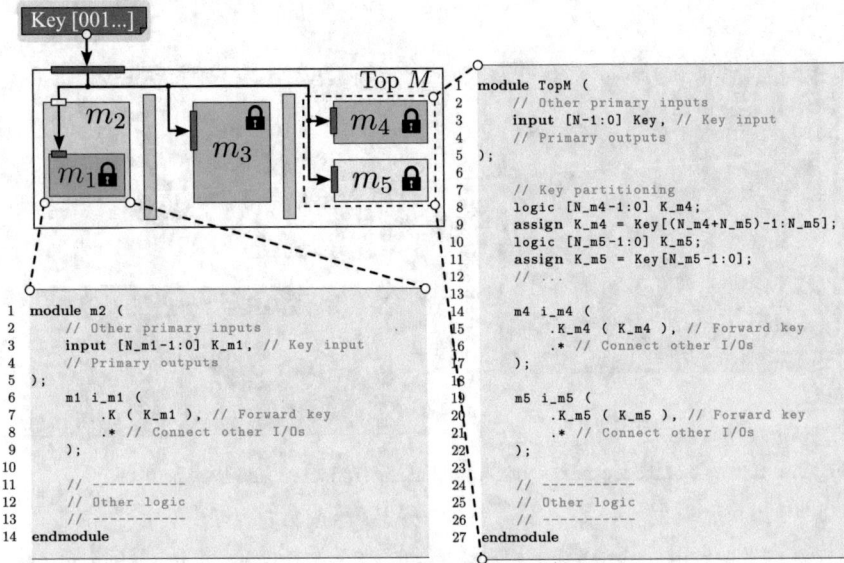

Fig. 7.7 The concept of routing the key through the HW hierarchy

the same hierarchical level as the top module, the key can be directly routed to the module inputs, as visualized for the modules m_4 and m_5. If a sub-module acts as a host of a locked component, the respective key partition is forwarded, as exemplified in the case of module m_2. The integration process finalizes the preparation of the locked HW design for the final stage—testing and verification.

7.6 Testing and Verification

At this point in the framework, the locked design is ready for the testing and verification process. This ensures that the previous stages did not induce functional errors and that a correct key results in a functionally equivalent HW design compared to the input RTL. Typically, this stage is performed in three steps: (1) key integration, (2) functional testing, and (3) equivalence checking.

7.6.1 Key Integration

To perform testing or verification, the correct activation key must be provided to the design. Depending on the use case, the key can be fetched from memory, provided via the verification tool, or, for simplicity, directly embedded into the code.

7.6.2 Functional Testing

To ensure a functionally correct behavior in the presence of the key, the locked HW can be tested using a dedicated benchmark suite. For example, in terms of processors, the testing can include executing a set of programs to cover all instruction types via simulation or FPGA-based emulation.

7.6.3 Equivalence Checking

Finally, to mathematically prove the equivalence between the RTL and the locked design, a formal verification tool can be deployed. In this book, we rely on using Synopsys Formality [180]. This tool is able to mathematically prove the functional equivalence of the two input files, regardless of their representation. The completion of this stage marks the final step in the framework. At this point, the HW design can proceed in the ASIC or FPGA flow.

7.6.4 Verification from RTL to Layout

As discussed in Sect. 4.4, the attack model can include an external design house to perform the layout generation. Regardless of its trustworthiness, an external design house typically has to perform verification steps to ensure that the layout tool did not introduce errors in the design. Not having access to the key, however, brings no disadvantage in this process. To ensure formal equivalence between RTL (in-house) and layout (external), the procedure is performed in two steps, as shown in Fig. 7.8. First, an in-house comparison between the RTL and the locked design with the correct activation key in place is performed (Fig. 7.8a). Second, the external design house compares the locked design (without the key) and the layout (Fig. 7.8b). If both steps conclude successfully, the transitive relation ensures that the RTL is equivalent to the final layout. Therefore, any errors induced by the synthesis and layout tools can be uncovered through verification even in the presence of logic locking and the inclusion of an external design house.

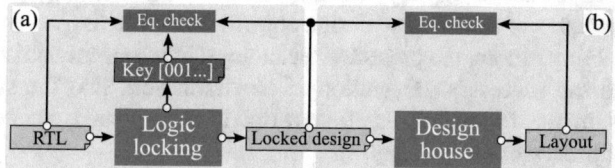

Fig. 7.8 Equivalence checking from RTL to layout: (**a**) in-house with the activation key (trusted) and (**b**) at the external design house without the key (untrusted)

7.6.5   Netlist Sign-off

Before exiting the trusted environment, the final locked design needs to be re-mapped to a specific target technology. Hereby, it is crucial to ensure the flattening of all design modules to remove any apparent structurally divisible components. This part is done separately depending on the specifics of the ASIC and FPGA flow.

7.7   Limitations and Outlook

The implementation of the framework has been designed to accommodate the untrusted design house and foundry attack model. In future studies, the framework can be extended to deal with threats posed by Hardware Trojan (HT)-infected third-party Intellectual Property (IP) and malicious EDA tools. The former is typically addressed through a set of detection and verification procedures to evaluate the absence of HTs and enforce specific security policies. The latter can be addressed via trustworthy formal verification tools or an output comparison between tools of multiple vendors.

Moreover, nowadays, the design of modern processor cores begins earlier in the IC design flow. Tools, such as the Synopsys ASIP Designer [178], offer the possibility to design application-specific processor architectures using more abstract language constructs—even before the RTL code is available. The description at higher abstraction levels allows for the simultaneous generation of the simulation and compiler tool chain as well as the target RTL code. Therefore, the output of these tools provides the initial input to the presented logic locking framework, thereby extending the window for potential integrity manipulations. Consequently, introducing security properties at higher design levels as well as ensuring the trustworthiness of the tools offers a potentially fruitful research direction.

7.8   Synopsis

This chapter presented an end-to-end framework for the application of logic locking schemes to multi-module HW designs. The framework components have been designed to adjust to the rapidly changing field of logic locking design and to enable the inclusion of security and design objectives early in the process of scheme deployment. Furthermore, the presented technology-independent netlist format has demonstrated the successful integration of the framework into the standard HW design flow. In the following, the framework is utilized and extended for the protection of processor cores (Chap. 8).

Chapter 8
Processor Integrity Protection

This chapter extends the framework introduced in Chap. 7 to protect multi-module Hardware (HW) designs—specifically processor cores—driven by two motivators. First, complex HW designs contain a multitude of interacting components. This native interaction offers the potential to scale logic locking across module boundaries, thereby offering another protective layer to counteract the Reverse Engineering (RE) process. Second, we recognize the importance of protecting critical in-processor signals against software-controller Hardware Trojans (HTs). In processor designs, these signals can easily be exploited to trigger controllable yet easy-to-implement HTs, which can compromise a system via a Denial of Service (DoS) attack. This chapter introduces the following:

- The design, implementation, and evaluation of a cross-module locking methodology for scaling logic locking beyond module boundaries [193, 195, 231]
- The design and evaluation of an inter-module encryption scheme for the protection of critical in-processor signals [194]

The rest of this chapter is organized as follows. Section 8.1 introduces the design and evaluation of the cross-module scheme on a RISC-V processor. Furthermore, within the same section, we showcase the successful application of the framework and the introduced scheme for the protection of the Made in Germany RISC-V (MiG-V) core. The concept of protecting processors against software-controller HTs is elaborated in Sect. 8.2. Section 8.3 discusses the limitations and outlook. A summary of this chapter is presented in Sect. 8.4.

8.1 Scaling Logic Locking Beyond Module Boundaries

So far, we have introduced a path toward locking complex HW designs, thereby driving the application of the protection policies based on specific security and

© The Author(s), under exclusive license to Springer Nature Switzerland AG 2023
D. Sisejkovic, R. Leupers, *Logic Locking*,
https://doi.org/10.1007/978-3-031-19123-7_8

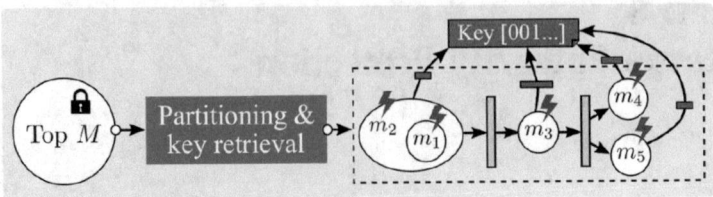

Fig. 8.1 Key-retrieval attacks through design partitioning

design objectives of the respective design components. In this section, the framework is extended to enable a locking mechanism across module boundaries by creating dedicated security dependencies. The motivation for this chapter is twofold. (1) As detailed in Sect. 4.8, the success of state-of-the-art RE methodologies depends on performing HW partitioning. This process subdivides the retrieved netlist into its structurally independent components. Only afterward, the identification of the components' functionality is performed. Therefore, it is critical to make the partitioning process difficult to execute in order to protect the design's secrets. (2) If every subcomponent of the target HW is locked individually, key-retrieval attacks can be deployed subsequently on every component. This is specifically effective in oracle-guided scenarios.

The two discussed aspects jointly create a path toward attacking complex locked HW as visualized in Fig. 8.1. Here, after partitioning, an attack on each module can be performed, yielding the complete key, assuming the attacks are successful. Thus, it is necessary to create locking-induced dependencies across modules to shift the attack focus from the subcomponents to the complete design. To address the motivating factors, we introduce Inter-Lock, a cross-module, scheme-independent locking procedure that creates security dependencies between selected components of the design. Herewith, the functionality of each interconnected HW component is adjusted according to an interdependency graph, yielding two outcomes. First, the selected modules become functionally dependent and structurally connected to make the partitioning process more complex. Second, the activation of each component relies on the activation of codependent modules. This forces an attacker to look at multiple components simultaneously to infer a correct key. Note that the effectiveness of Inter-Lock against a partitioning process is part of ongoing research activities, as successfully performing RE is a laborious and lengthy task [22, 198].

The rest of this section is organized as follows. Section 8.1.1 details the extensions that enable the application of Inter-Lock to complex HW designs. A security–cost trade-off of the proposed methodology on a 64-bit RISC-V core is presented in Sect. 8.1.2. Finally, Sect. 8.1.3 puts the framework and Inter-Lock into practice by introducing the first, fully logic-locked processor available on the market—the MiG-V.

8.1.1 Framework Extension: Introducing Security Dependencies

The core idea of Inter-Lock is to exploit the inherent multi-module nature of common HW designs; separate modules mutually depend on each other in terms of functionality and structure. This mutual dependence can be amplified to strengthen the functional and structural conjunction between all modules in a design. To utilize this HW feature for security purposes in addition to locking, Inter-Lock adapts the original design functionality of all selected modules to generate a *subset* of the activation keys for other, codependent modules during the activation procedure. This adaptation creates a security dependency between multiple components as follows: only if one module is correctly activated, its codependent modules can be unlocked as well. This security dependence has the following implications:

- Part of each module key is internally derived only after activation. Therefore, part of the key is hidden from the attacker.
- To fully activate the locked design, all components of the HW must be correctly activated at the same time. A single incorrect key in one module creates a chain reaction of functional failures across all security-dependent components.
- The additional security dependencies create functional inter-module connections that lead to higher congestion between the modules, counteracting the netlist partitioning process.
- As part of the key is embedded in the inherent functionality of the modules, to extract the complete key in an oracle-guided scenario, the adversary is forced to look at all connected modules at once.[1]

The overall Inter-Lock flow is presented in Fig. 8.2. The input consists of the generic netlists generated as output of the module preprocessing stage of the framework, as discussed in Sect. 7.3. The flow is divided into three steps: (1) First, the constellation selection assembles an interdependency graph between the selected modules (Fig. 8.2a). This graph defines the desired inter-module security dependencies. (2) Second, a selected logic locking scheme is applied on module level (Fig. 8.2b). (3) Finally, the locked designs are adjusted through the interlocking process to enable the security relations defined in the constellation (Fig. 8.2c). All steps are presented in more detail in the following.

---

[1] The reasoning behind this is based on the assumed attack model as defined in Sect. 4.9. As we assume that logic locking is effective in a low-volume production setting, where an activated Integrated Circuit (IC) is not available, the adversary needs to match the correct partitions against a database of activated circuits to perform oracle-guided attacks. However, if a cross-module scheme is deployed, the modules not only differ due to locking but also include additional functionality that creates a stronger connection to other components and infers additional inputs and outputs. Consequently, an exact matching can only be achieved at the top level of the design.

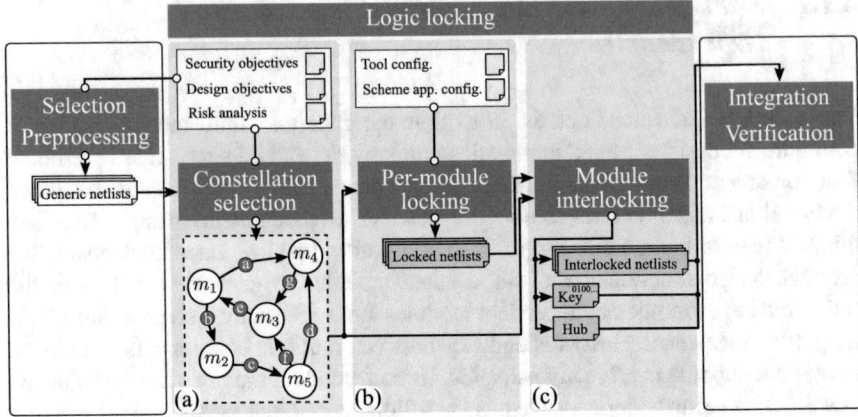

Fig. 8.2 The Inter-Lock flow: (**a**) constellation selection, (**b**) per-module locking, and (**c**) module interlocking

8.1.1.1 Constellation Selection

Constellations represent the selected structures in which the chosen modules are connected. The simplest way to visualize a constellation is through a weighted and directed dependency graph, as shown in Fig. 8.3a. Here, nodes represent modules, arrows indicate the security dependencies, and the weights define the dependency impact. The security dependency is defined by a *source* and a *sink* module, *where the activation of the sink depends on the activation of the source.* Consequently, only once a source module is correctly activated is it able to generate the correct *internal keys* to drive the activation of its sink modules. Internal keys are implemented as additional wires between the connected modules. The dependency impact defines how many internal key bits are generated between the source and the sink. Thus, it is important to differentiate external and internal key bits. The external key is derived from the memory and drives the primary key inputs of the top design as discussed in Sect. 7.5. The internal key bits are only set to their intended values after the external key is provided and all modules undergo a correct activation sequence.

A visual example of the described setup is provided in Fig. 8.3b. Here, the modules m_3 and m_4 are connected via two security dependencies, transforming both modules into a sink and a source. Note that the process of implementing security dependencies between modules is referred to as *interlocking*. The external activation keys for m_3 and m_4 are K_1 and K_4, respectively. m_3 generates the internal key K_3 for m_4, where $|K_3| = e$, while m_4 yields the internal key K_2 for m_3, where $|K_2| = f$. Therefore, the total correct keys for m_3 and m_4 are $K_1 \cup K_2$ and $K_3 \cup K_4$, respectively. As there is a mutual dependency between the modules, the overall correct behavior can only be extracted and compared against a golden model at the top-module level, not at the level of the individual components. Thereby, instead of attacking each component individually ($|K_1|$ and $|K_4|$), the attack vector moves to the top-module key ($|K_1| + |K_4|$). In this book, we refer to the total external

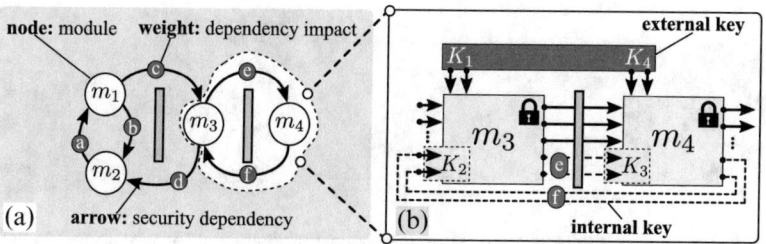

Fig. 8.3 Example of a constellation: (**a**) dependency graph and (**b**) visual example

key defined by Inter-Lock as *the effective key*. In the general case, the effective key length is defined as $K_e = \bigcup_{i=1}^{N} K_i$, where K_i is the key of module i, and N is the total number of codependent modules in the constellation. Thus, its length is defined as $\sum_{i=1}^{N} |K_i|$. Note that a correctly set external key for one module does not guarantee correct operation for the rest of the modules, as the external key inputs must be simultaneously correctly set.

Constellation Design In principle, there is no limitation to how a constellation is designed. However, every dependency leads to area, power, and performance overheads. Therefore, it is necessary to select a suitable constellation with respect to the security objectives and the available cost budget. As every HW design exhibits specific features, it is difficult to compile a clear set of design recommendations. However, we have identified a few simple rules to protect processor designs (following Fig. 8.3):

- It is advantageous to create more dependencies between modules that are close to each other, e.g., in terms of pipeline stages (e.g., m_3 and m_4). Otherwise, the delay overhead of the implementation can significantly rise.
- Registers should be placed between modules that are naturally divided by a pipeline stage (e.g., m_1 and m_3).
- Register placement is not required if the source and sink modules are located in the same pipeline stage (e.g., m_1 and m_2). However, potential combinational cycles must be resolved via register placement.
- Registers are not required if the source and the sink are in different pipeline stages and the dependency creates a forwarding path (e.g., m_4 to m_3).

Moreover, to ensure that an adversary has to consider all modules at once in an attack, all nodes must be included in the dependency graph. This can be done by designing a constellation in which every node has at least one input and one output dependency. Consequently, starting from any node, a dependency chain can be traced back to all other nodes in the constellation.

Extending the Application Setup To allow for a simple configuration of constellations in the framework, the application setup (Sect. 7.4.2.2) is extended as exemplified in Fig. 8.4a. The key lengths are set to small values to simplify the discussion.

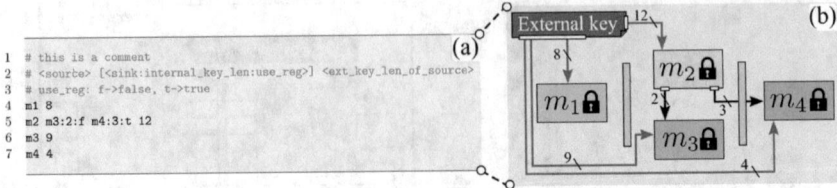

```
1  # this is a comment
2  # <source> [<sink:internal_key_len:use_reg>] <ext_key_len_of_source>
3  # use_reg: f->false, t->true
4  m1 8
5  m2 m3:2:f m4:3:t 12
6  m3 9
7  m4 4
```

Fig. 8.4 Example of a constellation setup: (**a**) application setup and (**b**) visual representation of the constellation

The extension allows an additional list of entries to be added to the setup in the form of `<sink:internal_key_len:use_reg>`. These entries define how a source impacts a sink module, including the number of generated internal key bits and a Boolean flag indicating if registers should be placed between the sink and the source for the internal keys. A visualization of the setup is presented in Fig. 8.4b. This extension has no impact on the regular application flow of the framework as the list of sink entries can be removed, degrading the setup to its standard form without Inter-Lock implications. Thus, some modules do not have to be involved in the interlocking procedure (see configuration for m_1).

Key Length Calculation It is important to notice that Inter-Lock is fully decoupled from the algorithmic specifics of locking schemes. The only interface between Inter-Lock and the selected scheme (deployed in the next step) is the total key length per module. As mentioned, first, this length is calculated based on the constellation. Afterward, each module is locked for a key of the given length. Hereby, part of the key is referenced from the external inputs and part is internally derived. The external per-module key length is defined by the parameters given in the application setup. The internal key length is defined as the sum of ingoing dependency impact weights. For example, in Fig. 8.4b, the total key provided to m_4 has 7 bits (4 external and 3 internal).

8.1.1.2   Per-Module Locking

The second step of the Inter-Lock flow concerns the application of a selected logic locking scheme, as shown in Fig. 8.2b. At this point in the flow, the exact key length for each module can be derived based on the constellation. Therefore, the framework randomly generates keys of the appropriate length and provides them to the locking scheme object. Due to the Object-Oriented Programming (OOP) nature of the design, the scheme object's locking function is called upon each module object, thereby providing the desired key as input. The output of the function is an object that encapsulates the concept of a key. This object contains ownership information about each bit of the key, including the separation of the total key based on the constellation. Therefore, the correct internal and external keys for each module can easily be extracted if necessary. The module object itself is only internally adjusted according to the mechanisms of the scheme. The configuration of the per-module scheme properties is done via the tool setup, as discussed in Sect. 7.4.2.1.

8.1.1.3 Module Interlocking

Once all modules are locked, the flow continues to the module interlocking step, as shown in Fig. 8.2c. The input to this step is a set of locked modules. The only missing part is the functional generation of internal keys to enable a correct activation chain. Hereby, the concept is as follows. Once activated, a selected source module must produce *correct* and *constant* (internal) output keys for all its sink modules. This behavior, however, must only hold true if the correct total input key (external and internal) is provided to the source. For all incorrect key values, the behavior of the generated keys must depend on the inputs provided to the design. To enable this functionality, the interlocking procedure integrates an Inter-Locking Circuit (ILC) into each source module. The properties of the circuitry are as follows: (1) the ILC generates correct key values only if the source module is correctly activated and (2) the ILC is structurally indistinguishable from the rest of the source implementation, i.e., a removal attack must be prevented. Both features are achieved by constructing the functionality of ILCs to be functionally and structurally dependent on the rest of the source module.[2] As presented in Fig. 8.5, the construction can be separated into three steps as follows.

Truth Table Construction First, a truth table is assembled based on the following input parameters (Fig. 8.5a): a set of wires from the original netlist ($\{w_1\}$), a set of

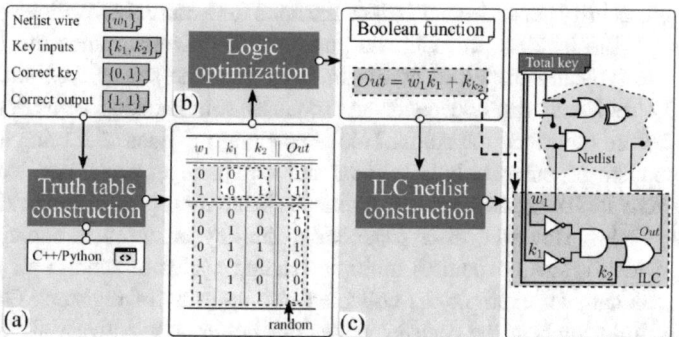

Fig. 8.5 Module interlocking flow: (**a**) truth table construction, (**b**) logic minimization, and (**c**) ILC netlist construction

---

[2] The functional dependency indicates that the output of ILC is influenced by the functional behavior of the source module. The structural dependency is a result of the functional dependency; the ILC is physically wired to the source module.

key inputs ($\{k_1, k_2\}$), the correct key assignment ($\{0, 1\}$), and the expected correct output ($\{1, 1\}$). The selected sets are exemplary, and their size can be adjusted as needed. The input-to-output assignment of the truth table reflects two cases: (1) when the correct key is given, the output is set to the expected value, and (2) for all other inputs, a random (or heuristically selected) output is assigned.

Logic Optimization Second, based on the constructed truth table, logic optimization is performed to minimize the implementation cost (Fig. 8.5b). The output of this step is a Boolean expression for the selected ILC. The open-source PyEDA library [52] is utilized for both the truth table construction and the optimization.

ILC Netlist Construction Finally, the received Boolean expression is mapped to a netlist by transforming each literal into a gate object (Fig. 8.5c). Based on the provided input wires and key inputs, the ILC netlist is embedded into the structure of the initial netlist.

The ILC construction principle ensures that an attacker cannot perform a simple removal of the circuitry. A removal attack would entail a correct isolation of the ILC logic from the initial netlist. As the circuitry has a functional dependency on the internal wiring of the design as well as on a set of key inputs, the identification of the logic is equivalent to the identification of any part of the locked circuit. However, the generic implementation allows for future adaptations to overcome potential security vulnerabilities. Whether structural isolation is possible is still part of active research.

8.1.1.4   Activation Procedure

The created security dependencies induce the need for a correct activation sequence. Even though the locking procedure is purely combinational, since the involved modules are typically dispersed across multiple pipeline stages, i.e., structurally divided by registers, the codependency requires simultaneous activation of all modules before the execution starts. This goes along the lines of forcing an attack on all modules at once; only a correct and complete external key leads to a correctly activated design. Otherwise, erroneous behavior is propagated in all further execution cycles. However, after production, the key is stored in some form of memory, which typically requires multiple reading cycles to acquire all key bits. The key bits can, for example, be collected into an array of registers. Only once the key is fully fetched, the register values are buffered to activate all necessary components. To enable a controlled activation in processor designs, the core's *reset* signal must be enabled until a correct key is loaded. Afterward, the reset is lifted and the core starts executing. Depending on the implementation of the ILCs, the activation might fail if not facilitated with multiple "free" cycles. In this context, two activation strategies are discussed as follows.

Cycle-Preventive Insertion To understand this scenario, let us assume that two modules impact each other, i.e., both act as source and sink in a closed interdependency loop. This constellation can create hazardous combinational loops, leading to an unstable activation sequence. This behavior can occur as both modules are never fully activated, i.e., one is waiting for the other, and vice versa. To prevent this cycle, the generated keys of the first module must not be placed in the input cone of the ILC of the second module. On the one hand, this cycle prevention is impairing the unrestricted dissemination of key gates in the design. On the other hand, this type of insertion does not require register placement. In both cases, if a longer chain of security dependencies exists, the processor core requires multiple cycles until a stable state is reached. Nevertheless, combinational cycles are typically avoided through register insertion or the design of the constellation itself.

Exclusive Insertion The second insertion type adapts the ILC implementation to depend only on external keys. Consequently, the activation itself requires only one cycle to propagate all necessary values. As the activation of the entire module still depends on the activation of its codependent components, no advantage is given to an adversary to isolate the ILCs. In both insertion variants, a correct reset function must exist to isolate the functional implication of the locked component until a stable state is reached. For example, the execution of a locked processor in a System-on-Chip (SoC) must be blocked to avoid faulty instruction execution until the activation is performed.

8.1.1.5 Hub Generation

In case a complex constellation is selected, it becomes a challenge to perform the key integration as the effective key must be correctly split and assigned to the respective modules through a correct wiring procedure. To simplify this process, the framework offers the option to generate a wiring *hub*. The core idea of the hub is to collect all external and internal keys and to automatically perform the key splitting and internal key routing. The hub can be generated based on the provided setup file. Verilog code examples of the hub generation are available in Appendix B.3.

8.1.2 Case Study: Protecting a RISC-V Core

This section evaluates the trade-off between the implementation cost and the effective key of the Inter-Lock methodology through a case study based on a processor design.

8.1.2.1   Experimental Environment

The evaluation was performed on the open-source, 64-bit RISC-V Ariane processor core [225]. To overcome the complexity of selecting all necessary Inter-Lock properties, e.g., constellation construction, the number of internal keys, key size, and others, the evaluation was performed according to preset rules. A set of effective keys of a fixed length was selected, including 1024-, 2048-, 2096-, and 8192-bit keys. Thus, the effective key was split among the selected modules linearly to the module size.

The Random Logic Locking (RLL) scheme was selected to perform the per-module locking. The design of RLL is beneficial for evaluation, as its cost is linear to the key length (up to two gates per key bit), and the key gates are disseminated randomly across the design, i.e., without making a biased decision toward a more cost-friendly insertion. Moreover, as discussed in Chap. 5, RLL presents a superset of a large class of schemes, as other methods are built upon the concept of XOR/XNOR gates, only with a more selective choice of suitable insertion locations.

The Synopsys Design Compiler (DC) [179] was used for logic synthesis with the standard-performance cell library for the UMC 90 nm CMOS process, operating under typical conditions (1 V, 25°C). The Register-Transfer Level (RTL) and gate-level simulations were performed using ModelSim [161]. Formal verification was done with Formality [180].

8.1.2.2   Constellation Design

For the purpose of this case study, a set of critical design modules of the processor core were extracted for locking. These are presented in Table 8.1. The selected modules were processed through all stages of the presented framework from Chap. 7. The constellation was constructed based on the position of all modules

Table 8.1 Ariane combinational modules

Module	Abbreviation	#inputs	#gates	#outputs
flush_controller_logic	f_ctrl	145	22	11
csr_buffer_logic	c_buff	222	134	90
instruction_scan	i_scan	32	240	139
instruction_realigner_logic	i_real	183	629	276
compressed_decoder	c_dec	32	848	34
commit_stage	commit	985	1584	417
branch_unit	br_unit	342	1655	328
decoder	decoder	518	2169	362
pc_select	pc_sel	521	3333	128
branch_prediction	br_pred	814	4669	333
alu	alu	206	7412	65

Fig. 8.6 Selected constellation for the Ariane core

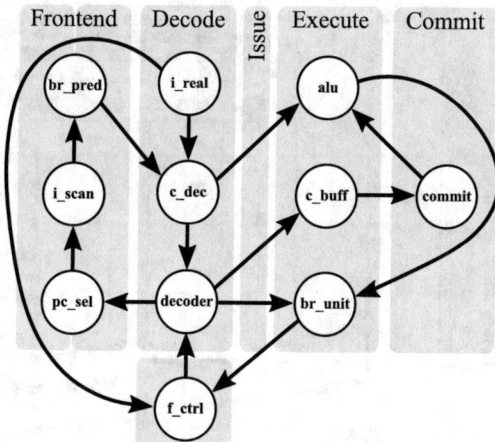

in the core's pipeline. The register insertion was done using exclusive insertion. As a rule of thumb, the total number of internal wires generated by a source module was set to 5% of the number of its original outputs. In the case of multiple sinks for a single source, the generated internal keys were evenly divided among the components. The final constellation is presented in Fig. 8.6. Here, the module nodes are visualized in their respective pipeline stage. The constellation was selected based on the recommendations from Sect. 8.1.1.1. Note that each node in the dependency acts as a source and a sink, thereby ensuring a closed dependency chain.

8.1.2.3 Cost Evaluation

Before further discussions, it is important to summarize the security aspects of Inter-Lock. As this meta-scheme is independent of the actual locking policy, its security is mainly defined within the frame of the security properties of the underlying locking scheme. However, due to the assumed attack model, we approximate the security based on the effective key length. Thus, the presented evaluation focuses on the cost impact of the various effective key lengths on the locked design.

The analysis is based on a well-proven evaluation tool in hardware design—the Area-Timing (AT) plot. The AT curve visualizes the design cost for a range of clock periods (T_{clk}). The area is represented in Gate Equivalent (GE) and the clock period in ns. More details on the concept of the AT plot can be found in Appendix D.2.

The locked Ariane core variants were evaluated for a range of clock periods with a step of 0.25 ns, starting from $T_{clk} = 7$ ns until the lowest viable T_{clk} was reached. The lowest T_{clk} is constrained by the critical path of the design, while the highest T_{clk} is manually selected to a sufficiently high value to display a reasonable AT curvature. The retrieved AT plot is shown in Fig. 8.7a. The optimum AT points for the original design and all locked variants are highlighted in the shown plot. Note that the evaluation excluded the memory cells, as these have not been affected

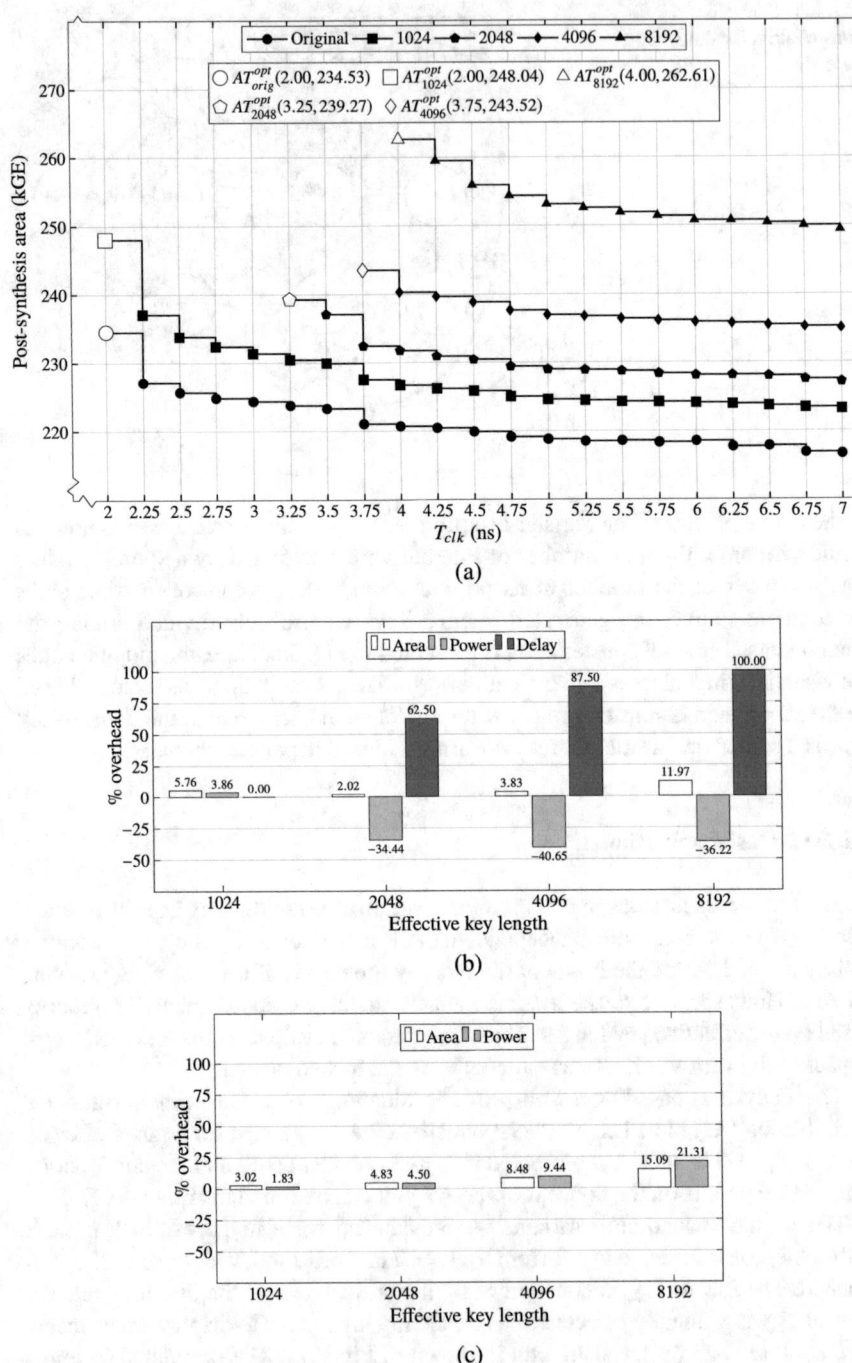

Fig. 8.7 Inter-Lock security–cost evaluation. (**a**) Area-timing plot. (**b**) Optimum AT and high performance. (**c**) Low performance

by logic locking. To extract a realistic cost, the AT results can be compared at different comparison points, including: optimum AT, high performance, and low performance.

Optimum AT and High Performance   The area, power, and delay overhead of the locked variants are presented in Fig. 8.7b. In this evaluation, the AT optima correspond to the high-performance comparison points, i.e., to the shortest achieved clock periods. Therefore, the following observations hold for both cases. The area overhead across all key lengths ranges from 2.02% to 11.97%. Hereby, the area overhead is lower for the 2048-bit and the 4096-bit key, compared to the 1024-bit and 8192-bit keys. The reason is likely due to the following. First, the 1024-bit key has a 0% delay overhead compared to the original AT optimum, i.e., the locking mechanism has not impacted the critical path. However, to reach the same performance, the design area rises to accommodate faster design implementations. As the 1024-bit key reaches a significantly shorter clock period, it is expected that its area overhead is higher compared to the 2048-bit and 4096-bit keys. Second, the 8192-bit key introduces a much higher number of key gates, resulting in the highest area overhead. A high number of key gates is more likely to affect the critical path. Thus, here, the delay overhead reaches 100% at $T_{clk} = 4$ ns. However, as the locked design itself starts at the highest gate count even for long clock periods (low performance) compared to all other keys, due to the AT curvature, the high-performance point for the 8192-bit key leads to the highest cost impact and slowest design. In terms of power, the overhead decreases for larger keys as the comparison points implement significantly slower designs. Note that the power overhead is approximated with Synopsys DC.

Low Performance   The comparison at low-performance points looks at the case when no excessive optimizations have been introduced yet in the synthesis process. Therefore, the AT curve is typically leveled out for long clock periods, yielding a fair comparison. For this evaluation, the designs at $T_{clk} = 7$ ns (142.85 MHz) were selected. Note that any clock period toward the right end of the AT curve would yield similar results. The area and power overheads for all key variants are presented in Fig. 8.7c. Due to the linear cost nature of the selected per-module locking scheme, the area overhead is linearly increasing (approximately with a factor of up to 2x) with the key length, ranging from 3.02% to 15.09%. The same conclusion can be drawn for the power overhead, which correlates with the area increase due to the fixed clock period. This evaluation indicates that the cost of Inter-Lock is highly dependent on the selected locking scheme, rather than the included ILCs and necessary wiring.

8.1.2.4   Testing and Verification

To verify that Inter-Lock did not introduce functional faults in the locked design, functional testing and formal verification were deployed, based on the discussion in Sect. 7.6. The functional testing was performed using the available open-source RISC-V test suite, containing a multitude of assembly and benchmark tests [189, 225].

8.1.2.5 Summary

The presented evaluation provides a closer look at the cost of deploying locking schemes to a practical, silicon-proven processor design at a larger scale, thereby considering interdependencies between multiple design modules and large external keys. Moreover, based on the evaluation, a protected processor design can be chosen by balancing the area, power, and delay overhead against the effective key length.

8.1.3 The "Made in Germany RISC-V" Core

The contributions of this book have been continuously channeled into practical applications through the collaboration with HENSOLDT Cyber GmbH—a corporate startup driving the future of trustworthy systems, in both the hardware and software domains. Through joint efforts, the framework presented in Chap. 7 and its Inter-Lock extension have been successfully transferred to HENSOLDT Cyber GmbH, resulting in the production of the Made in Germany RISC-V (MiG-V)—the first fully logic-locked, general-purpose processor available on the market [70]. A snapshot of the MiG-V is shown in Fig. 8.8a.[3]

The processor is based on the open-source, 64-bit, 6-stage, in-order, single-issue Ariane core [225], a microarchitectural implementation of the open-source RISC-V Instruction Set Architecture (ISA) [197]. Specifically, the core implements the

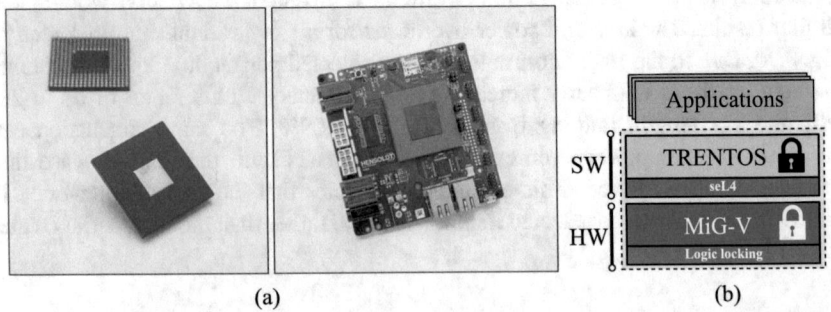

(a)   (b)

Fig. 8.8 Logic locking in practice: (**a**) the protected MiG-V core and (**b**) a secure HW–SW solution

---

[3] The MiG-V images have been provided by HENSOLDT Cyber GmbH.

RV64IMAC ISA, including 64-bit integer operations with standard extensions for the integer multiplication and division (M), as well as atomic (A) and compressed instructions (C). The MiG-V has been fabricated using a 55 nm Complementary Metal–Oxide Semiconductor (CMOS) process.

In terms of integrity protection, MiG-V is locked with a 1024-bit key through Inter-Lock. Hereby, the attack model follows the assumptions of this book, protection against intelligible hardware Trojans in the processor core. For the first generation of the chip, the attack model does not extend beyond the limits of the untrustworthy external design house and foundry. Thus, the main attack vector is defined by an oracle-less-based reverse engineering process, as discussed in Sect. 4.9.

The MiG-V is designed for deployment in security-critical applications. To extend the security objectives beyond hardware, the Trusted Entity Operating System (TRENTOS) [71] is placed on top of the core, as visualized in Fig. 8.8b. TRENTOS is based on seL4 [68]—a formally verified microkernel. The core feature of seL4 is the assurance of isolation between applications running in the system. Therefore, a potential attack in one application can be *contained*, preventing the attack from compromising other system components. This unique composition of the locked MiG-V and the secure operating system enables a secure hardware–software solution from the ground up. More details on the core's key features can be found in [69, 70, 195].

8.1.4   Security Analysis

The security aspects of Inter-Lock are twofold, including the security of the per-module scheme and the security of the interlocking procedure. Modeling the exact security level of Inter-Lock is difficult as its security depends on the success of the reverse engineering process and the isolation of the ILC components, hence transcending the traditional concepts of logic locking. However, the effective key length provided by Inter-Lock can capture the security aspects of specific locking schemes under the assumed attack model, offering a simple measure of security. Nevertheless, Inter-Lock continues to be challenged for potential security vulnerabilities. A practical example is manifested in a red team versus blue team analysis, managed by the industrial partner HENSOLDT Cyber GmbH, to evaluate the security policies involved in the protection of the MiG-V. Moreover, as logic locking is rapidly evolving, it is crucial to enable a fast-tracked integration of scheme improvements for the next generation of secure hardware designs. Precisely for this reason, the framework's modular implementation provides a simple interface to address future vulnerabilities. At the time of writing this book, Inter-Lock has not been compromised [23].

8.2 Protecting Against Software-Controlled Hardware Trojans

Hardware Trojan design is driven by two motivators. First, the trigger of an HT must provide a controllable mechanism to activate the Trojan's payload. Second, the payload itself must lead to the manifestation of a predetermined attack. If both the trigger and the payload implementation leave a minimal footprint in the HW design, the HT is more likely to pass any detection mechanisms. In the context of processor designs, the attacker has two possible trigger sources: hardware and software. This extends the attacker's capabilities to implement simple yet controllable Trojans. Often, it is assumed that HTs are embedded inside a module. This assumption requires a full recovery of the activation key to perform the Trojan insertion. In this section, we relax the attack model by assuming the following scenario. By only identifying a few control signals indicating the instruction type between processor modules, an attacker is able to design a hardware Trojan that can exploit software-dependent signal behavior and launch a DoS attack resulting in a complete system shutdown.

Attack Model As a motivational example, let us consider the HT shown in Fig. 8.9. The Trojan is located between the decoder and the Arithmetic Logic Unit (ALU) of the 32-bit, 4-stage RI5CY processor core. The RI5CY core implements the open-source RISC-V ISA [197] within the PULPino SoC [188]. In this microarchitecture,

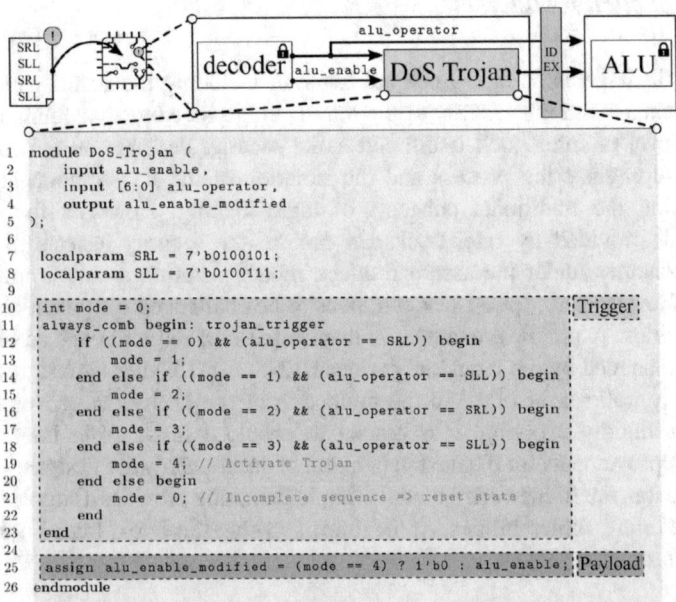

```
1  module DoS_Trojan (
2      input alu_enable,
3      input [6:0] alu_operator,
4      output alu_enable_modified
5  );
6
7  localparam SRL = 7'b0100101;
8  localparam SLL = 7'b0100111;
9
10 int mode = 0;                                                    Trigger
11 always_comb begin: trojan_trigger
12     if ((mode == 0) && (alu_operator == SRL)) begin
13         mode = 1;
14     end else if ((mode == 1) && (alu_operator == SLL)) begin
15         mode = 2;
16     end else if ((mode == 2) && (alu_operator == SRL)) begin
17         mode = 3;
18     end else if ((mode == 3) && (alu_operator == SLL)) begin
19         mode = 4; // Activate Trojan
20     end else begin
21         mode = 0; // Incomplete sequence => reset state
22     end
23 end
24
25 assign alu_enable_modified = (mode == 4) ? 1'b0 : alu_enable;    Payload
26 endmodule
```

Fig. 8.9 Denial of service hardware Trojan in a RISC-V core

the instruction `opcode` is represented in the form of the `alu_operator`, a 7-bit signal generated as output of the instruction decoder module. This signal is forwarded through pipeline registers from the Instruction Decode (ID) stage to the ALU in the Execution (EX) stage to control which operation will be executed in the next cycle. The DoS Trojan is designed based on the following scenario. The attacker is able to identify the `alu_operator` signal, even if the decoder and the ALU are locked. Using the exposed signal, the HT can detect the execution of a predefined, rare *sequence of instructions*. In the shown example, a sequence of logical left (SLL) and logical right (SRL) shifts is selected: SRL, SLL, SRL, SLL. If the sequence is detected, the HT is triggered, blocking the execution of all further ALU instructions—effectively performing a DoS attack. In the RI5CY core, the ALU is deactivated by setting the `alu_enable` signal to 0. This simple HT can lead to dangerous consequences, especially in the context of medical devices, and military and automotive electronics. We identified this form of software-controlled HTs as particularly alarming due to the following reasons:

- The Trojan can potentially be inserted without the need to fully unlock all modules in a design, as only a very small subset of signals need to be localized. This is further strengthened by the fact that the Trojan is based on the expected behavior of the inter-module signals *after activation*. It stands to reason that it is easier to identify inter-module control signals than internal signals within one module [176].
- The HT relies on the *expected* behavior of signals. Thus, even if logic locking ensures an incorrect behavior of the identified signals in the presence of an incorrect key, the HT will still perform correctly after the activation.
- The expected behavior of the identified signals is exposed in a freely available source, i.e., in the RISC-V ISA manual.
- Finally, controlling the Trojan via software enables a vast attack landscape of controllable attacks.

To address this attack scenario, in the following, we present the design and implementation concept of Control-Lock, an encryption methodology for the protection of inter-module control signals. The working principle of this scheme is to ensure that the selected signals behave incorrectly even when the correct keys are applied. Therefore, the HT is not able to exploit the expected behavior of the signals after activation. Consequently, the attacker is required to fully unlock all relevant modules before being able to understand the encrypted signal behavior.

The rest of this section is organized as follows. The design and implementation concepts of Control-Lock are presented in Sect. 8.2.1. A key-dependent netlist generation procedure is discussed in Sect. 8.2.2. Multiple encryption variants are presented in Sect. 8.2.3. The introduced methodology is evaluated on a RISC-V case study in Sect. 8.2.5. Related work is discussed in Sect. 8.2.6.

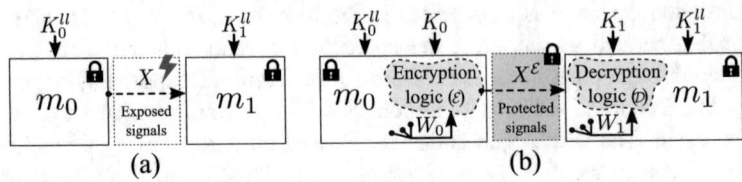

Fig. 8.10 Example of (**a**) exposed and (**b**) protected control signals

8.2.1 The Control-Lock Methodology

The overall concept of Control-Lock is presented in Fig. 8.10. Initially, the two selected modules are locked with the keys K_0^{ll} and K_1^{ll}, respectively. Here, module m_0 outputs a set of signals represented as $X = \{x_0, \ldots, x_{n-1}\}$. Every element $x_i \in X$, where $0 \leq i < |X|$, denotes a 1-bit signal, i.e., one output signal of m_0. In this setting, X acts as a set of control signals for module m_1. Without Control-Lock (Fig. 8.10a), X is exposed to an adversary even if the correct activation key is not found.

   To protect X, Control-Lock assembles a randomized encryption ($\mathcal{E}$) and decryption ($\mathcal{D}$) logic. As shown in Fig. 8.10b, the output signals X of module m_0 are modified by $\mathcal{E}$, generating the encrypted set of signals $X^{\mathcal{E}}$. These are forwarded to module m_1. Therefore, $\mathcal{E}$ is appended to module m_0. To be processed by m_1, $X^{\mathcal{E}}$ needs to be decrypted. This is handled by the decryption logic $\mathcal{D}$, which is integrated at the inputs of module m_1. As Control-Lock protects the control signals between m_0 and m_1, $X^{\mathcal{E}}$ cannot easily be utilized for Trojan insertion, even if identified.

8.2.1.1 Implementation Concept

$\mathcal{E}$ and $\mathcal{D}$ are manifested as key-dependent *sets of encodings* of X to $X^{\mathcal{E}}$ and vice versa. Thereby, every key value defines a randomized encoding and decoding pair. Only a correct activation key driving $\mathcal{E}$ and $\mathcal{D}$ results in a correct encoding and decoding sequence. $\mathcal{E}$ is represented as a netlist with three inputs: the original signals X, a set of key inputs $K_0 = \{k_0, \ldots, k_{j-1}\}$, and a set of randomly selected wires from the original netlist $W_0 = \{w_0, \ldots, w_{l-1}\}$. All elements of K_0 and W_0 are 1-bit signals. Thus, $\mathcal{E}$ defines the mapping of X to $X^{\mathcal{E}}$, based on K_0 and W_0 for module m_0:

$$\mathcal{E} : (K_0, W_0, X) \rightarrow X^{\mathcal{E}}. \tag{8.1}$$

Consequently, $\mathcal{D}$ maps the corrupted set $X^{\mathcal{E}}$ back to the original values of X, based on K_1 and W_1 for module m_1:

$$\mathcal{D} : (K_1, W_1, X^{\mathcal{E}}) \rightarrow X, \text{ where } K_1 \neq K_0. \tag{8.2}$$

Hereby, the selection process for W_1 must prevent combinational cycles in the design of $\mathcal{D}$. Preferably, the set contains wires that are taken near the netlist inputs of m_1 and not positioned in the output cone of $X^{\mathcal{E}}$.

8.2.1.2 Design Objectives

To maximize the security and applicability aspects of Control-Lock, the implementation is steered by the following design objectives: (1) key-dependent corruption, (2) indistinguishability of $\mathcal{E}$ and $\mathcal{D}$, and (3) compatibility with any logic locking scheme.

Key Dependency The design of Control-Lock ensures the dependency on a correct key. Correctly set keys imply that both $\mathcal{E}$ and $\mathcal{D}$ are able to encrypt and decrypt the selected signals correctly. If one of the keys K_0 or K_1, where $K_0 \neq K_1$, is incorrect, the encryption fails and *the overall design functionality* remains corrupted, as visualized in Fig. 8.10b.

Indistinguishability The indistinguishability property mitigates structural removal attacks on $\mathcal{E}$ and $\mathcal{D}$. This property is fulfilled based on the structural composition of $\mathcal{E}$ and $\mathcal{D}$; both components depend on a set of *original* wires W_0 and W_1, chosen from the initial netlists. For example, $\mathcal{E}$ in Fig. 8.10b depends on W_0. Thus, the functionality of $\mathcal{E}$ is driven by W_0, i.e., part of the original logic. After resynthesis, both $\mathcal{E}$ and $\mathcal{D}$ are structurally integrated into m_0 and m_1, respectively. Thereby, the original X is no longer identifiable as it becomes an *internal* module signal. Moreover, the key inputs of the logic locking mechanism and the Control-Lock components are merged, making a division of the two key sets difficult to perform.

Compatibility Since Control-Lock can be deployed before or after the logic locking procedure, the compatibility objective is fulfilled. Thus, the methodology does not affect the security of the selected logic locking policy.

8.2.2 Key-Dependent Netlist Generation

In the following, only the implementation of the encryption logic is exemplified, as the decryption counterpart is designed using the same principles. For correct keys, $\mathcal{E}$ and $\mathcal{D}$ must perform a valid encryption and decryption sequence. Otherwise, incorrect keys must result in the random behavior of $X^{\mathcal{E}}$. For this purpose, the key-dependent netlist generation flow is introduced. As shown in Fig. 8.11, the input of the flow consists of a set of signals X, a selected key K_m, and a set of randomly selected signals (wires) W_m from the module m. The core mechanism is based on the Control-Lock Circuit (CLC) construction, which consists of three steps: TT generation, logic optimization, and netlist integration. In principle, the flow is

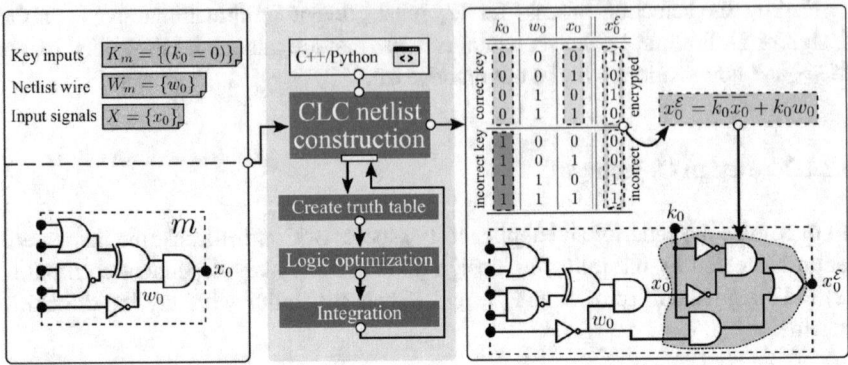

Fig. 8.11 The key-dependent netlist generation flow

equivalent to the ILC generation, described in Sect. 8.1.1.3. The only difference lies within the way the netlist is constructed based on the provided inputs.

The example in Fig. 8.11 showcases the truth table generation using only 1-bit input signals, i.e., where $|K_m| = 1$, $|W_m| = 1$, and $|X| = 1$. In the first step, the TT is initialized. The input of TT is defined with the group of signals (K_m, W_m, X). An incorrect output is assigned for every (K_m, W_m, X) pair for $k_0 = 1$. In the case of a correct key, i.e., $k_0 = 0$, a selected encoding is generated and inserted into TT for every $(K_m, -, X)$. Hereby, the encoding does not functionally depend on the values of W_m. Afterward, the TT is minimized, and the corresponding Boolean expression is extracted. Finally, the Boolean function is transformed to a netlist representation and embedded into the original design. The same concept is repeated for the generation of the decryption logic to reverse $x_0^{\mathcal{E}}$ back to x_0.

8.2.3 Signal Grouping Schemes

So far, every bit $x_i \in X$ has been encrypted and decrypted individually, as shown in Fig. 8.11. However, in order to achieve a larger encoding/decoding space for a given key, we introduce the concept of *signal grouping schemes*. A grouping scheme defines how X is divided into groups of signals, where all signals of the same group are jointly encrypted. A grouping scheme can be defined as follows:

$$\mathcal{E} = \left(\mathcal{E}_0^{N_0}, \dots, \mathcal{E}_{G-1}^{N_{G-1}} \right), \tag{8.3}$$

where X is divided into G groups and N_k is the number of bits in the group k, where $0 \leq k < G$ and $\sum_{k=0}^{G-1} N_k = |X|$.

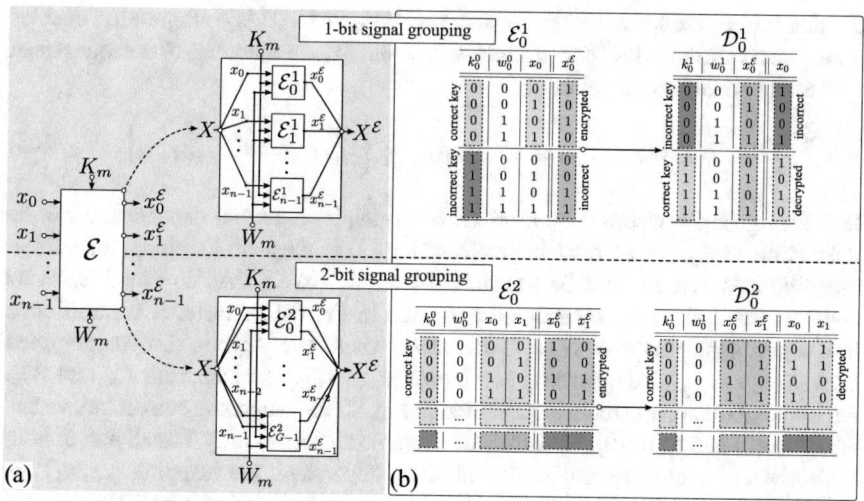

Fig. 8.12 Control-Lock 1-bit and 2-bit grouping scheme: (**a**) design concept and (**b**) implementation example

8.2.3.1   1-bit Grouping

Let us first consider the 1-bit grouping scheme where every $N_k = 1$, as shown in Fig. 8.12a. Here, $\mathcal{E}$ is implemented as the grouping scheme $\mathcal{E} = (\mathcal{E}_0^1, \ldots, \mathcal{E}_{G-1}^1)$. Every $\mathcal{E}_k^1$, where $0 \leq k < G$, defines the encryption of a single x_k individually:

$$\mathcal{E}_k^1 : (K_m, W_m, \{x_k\}) \rightarrow \{x_k^{\mathcal{E}}\}. \tag{8.4}$$

Thus, $G = |X|$ and $x_i = x_k$. This grouping scheme can be implemented as shown in Fig. 8.12b. Here, for every x_k, the pair $(\mathcal{E}_k^1, \mathcal{D}_k^1)$ is implemented with a separate TT. In the given example, we assume that the correct keys are $k_0^0 = \{0\}$ and $k_0^1 = \{1\}$, for $\mathcal{E}_0^1$ and $\mathcal{D}_0^1$, respectively. Apart from K_m and W_m, every $x_k^{\mathcal{E}}$ depends only on its original x_k. Thus, one TT per x_k is assembled. For a correct mapping, the information about how the encryption affects x_k must be transferred to the design of $\mathcal{D}_k^1$. For example, if a selected x_k is inverted by $\mathcal{E}_k^1$, then $\mathcal{D}_k^1$ has to invert it back to preserve the original value. Consequently, the 1-bit grouping scheme can only implement one functionality per x_k: the signal is either *buffered* or *inverted* for a correct key.

8.2.3.2   N-bit Grouping

To enable a stronger encryption, one $\mathcal{E}_k^{N_k}$ can jointly encrypt multiple x_k ($N_k > 1$) by creating a mutual dependency. To simplify the discussion, let us assume that X is divided into G groups of the same size, i.e., $\forall N_k = N$. This configuration enables a new encryption policy. For a selected K_m and W_m, a mapping of a sequence of values can be defined as follows:

$$\mathcal{E}_k^N : (K_m, W_m, \{x_{k \cdot N}, \ldots, x_{k \cdot N + (N-1)}\}) \rightarrow \left\{ x_{k \cdot N}^{\mathcal{E}}, \ldots, x_{k \cdot N + (N-1)}^{\mathcal{E}} \right\}, \qquad (8.5)$$

for every group k where $0 \leq k < G$. In principle, a mutual dependency can be created for any group of N elements from X. As for the 1-bit grouping, the correct mapping information must be transferred from $\mathcal{E}_k^N$ to $\mathcal{D}_k^N$. As an example, let us consider the 2-bit grouping scheme presented in Fig. 8.12a. Here, X is partitioned into subsequent groups of $N = 2$ elements, where every group of two encrypted signals $\{x_{k \cdot 2}^{\mathcal{E}}, x_{k \cdot 2 + 1}^{\mathcal{E}}\}$ depends on two inputs $\{x_{k \cdot 2}, x_{k \cdot 2 + 1}\}$, apart from K_m and W_m. An implementation example is presented in Fig. 8.12b, where the correct key values are $k_0^0 = \{0\}$ and $k_0^1 = \{0\}$. Note that the example visualizes only a small part of both truth tables. Because x_0 and x_1 are mutually dependent, the mapping $\{x_0, x_1\} \rightarrow \{x_0^{\mathcal{E}}, x_1^{\mathcal{E}}\}$ can be defined for all correct and incorrect keys. Afterward, $\mathcal{D}$ is required to map all encrypted values $\{x_0^{\mathcal{E}}, x_1^{\mathcal{E}}\}$ back to the original form $\{x_0, x_1\}$ for a correct key.

This joint encryption has a beneficial consequence: a single x_0 cannot be correctly decrypted without simultaneously considering the value of x_1 and vice versa. However, higher groupings are also costlier. Furthermore, $\mathcal{D}$ can have different implementations for the same $X^{\mathcal{E}}$ in the case when X is buffered as input to multiple separate modules, further facilitating the removal resilience of Control-Lock.

8.2.3.3   Framework Extension

The Control-Lock methodology has been implemented as a modular extension of the framework described in Chap. 7. The extension is plugged into the flow by replacing the logic locking stage of the framework (see Fig. 7.1c). The encryption and decryption logic is designed and implemented for a set of signals for two or more selected modules. Hereby, the same risk analysis and selection procedure is implied as for logic locking (Sect. 7.2). Similar to the implementation of ILCs, the PyEDA infrastructure is utilized for the truth table construction and optimization.

8.2.4   Security Analysis

The security of Control-Lock can be seen through two aspects: the *overall* and the *inherent* security. The overall security depends on the security of the selected logic locking scheme due to the following reasons. First, Control-Lock changes the original functionality of the selected modules, thereby tightly integrating the key-dependent mapping process into the design's structure. Second, the Control-Lock key is embedded into the logic locking key, making it indistinguishable from the rest of the locking mechanism. Therefore, the primary goal of an attacker is to first retrieve the logic locking key, before being able to reason about the encrypted signals. The inherent security aspects are driven by the dependency among signals (N-bit grouping) and the key length ($|K_m|$). Both factors exponentially increase the number of possible mappings from X to $X^{\mathcal{E}}$. Hereby, each distinct key value can activate a selected mapping, where the total number of mappings per key and group is equivalent to the number of permutations for a set of size 2^{N_k}. Therefore, as with all hardware protection policies, Control-Lock can be attacked from two perspectives: structure and functionality. In the former case, the correct locations (indexes) and the number of key bits must be identified. Afterward, a structural analysis must be able to successfully isolate the CLC from the rest of the design. This process is hampered due to the structural and functional dependency of CLCs on the original design. A functional attack is only viable after the involved modules have been unlocked, under the assumption of a correct key separation. Afterward, the correct mapping must be identified via a brute-force search of all available groupings and mappings.

In the DoS setting, even if an attacker is able to identify the correct signals without unlocking the design, a Trojan insertion remains infeasible, as the trigger cannot utilize the expected behavior of the signals. Thus, Control-Lock excludes the direct utilization of critical signals as Trojan triggers by forcing the attacker to first unlock the overall design.

8.2.5 Case Study: Protecting Against a Denial of Service Trojan

Since the CLC construction is highly dependent on the selected mapping for correct and incorrect keys as well as the size and grouping of X, it becomes a challenge to perform a generic cost evaluation. Therefore, the evaluation is focusing on the concrete Trojan scenario as described in Fig. 8.9. Hereby, the 7-bit `alu_operator` signal between the ALU and the decoder of the RI5CY core was selected for encryption with Control-Lock. Thus, the encryption unit $\mathcal{E}$ was integrated into the decoder, and the decryption unit $\mathcal{D}$ was embedded into the ALU.

8.2.5.1 Experimental Environment

The RI5CY core was used to evaluate the cost of Control-Lock. ModelSim [161] was utilized for the RTL and post-synthesis simulation. Logic synthesis was performed with Synopsys DC [179] and the UMC 90 nm CMOS technology library. Formal verification was done with Formality [180].

Table 8.2 Initial module
area, delay, and power values
for the RI5CY core

Module	Area (GE)	Min. delay (ns)	Power (mW)
Decoder	1300	0.25	1.59
ALU	9646	1.50	4.13
RI5CY core	47,000	2.25	300

To evaluate the cost impact of various grouping schemes, the 7-bit
alu_operator signal was evaluated for four different schemes in the form of
$(N_0, N_1, \cdots, N_{G-1})$, including: $(1, \cdots, 1)$, $(2, 2, 2, 1)$, $(2, 2, 3)$, and $(3, 4)$. Every
value in a scheme indicates one selected N_k-bit grouping, where $\sum_{k=0}^{G-1} N_k =$
$|alu\_operator|$. The initial area, delay, and power values of the selected modules
and the entire core are presented in Table 8.2. For simplicity, we assume $|W| = 1$
and $|K| = |X|$.

To evaluate the security–cost trade-off, an AT plot for both modules was
generated, and the area, power, and delay overheads for both modules and the
selected grouping schemes were extracted. Hereby, the evaluation focuses on three
comparison points: optimal AT, high performance, and low performance. However,
as all variants do not affect the critical path of the overall core, in the following, we
only discuss the *high-performance* results. Other evaluation results are presented in
Appendix D.3.

8.2.5.2   Results

The high-performance evaluation results are shown in Fig. 8.13. For both modules,
the high-performance points diverge in the achieved clock period. Thus, the area and
power overheads significantly vary across different schemes, making it difficult to
extract a common observation across all cost properties. This is specifically true for
the smaller decoder module, where the integrated CLCs yield a higher impact on the
module's implementation. Nevertheless, the following observations can be made.
For the small decoder module, the delay overhead increases with more complex
schemes, ranging from 0.00% to 100.00%. Consequently, the high-performance
points diverge in the achieved clock period, yielding an area overhead from −3.90%
to 8.44%, and a power overhead from −53.09% to 2.03%.

The overhead values for the larger ALU module are more limited in range as the
embedded CLCs have, in total, a lower impact on the design. Thus, the area overhead
varies from −2.64% to 12.72%, and the power overhead ranges from −9.53% to
3.74%, and the delay is limited to a maximum of 15.38% for the most complex
scheme.

The broad overhead range (especially for the smaller design) is exacerbated by
the randomness of the CLC construction. Nevertheless, in the context of the entire
RI5CY core, the area overhead implied by the most complex scheme is only 4.75%
without any performance degradation. Thus, the results indicate that it is favorable to

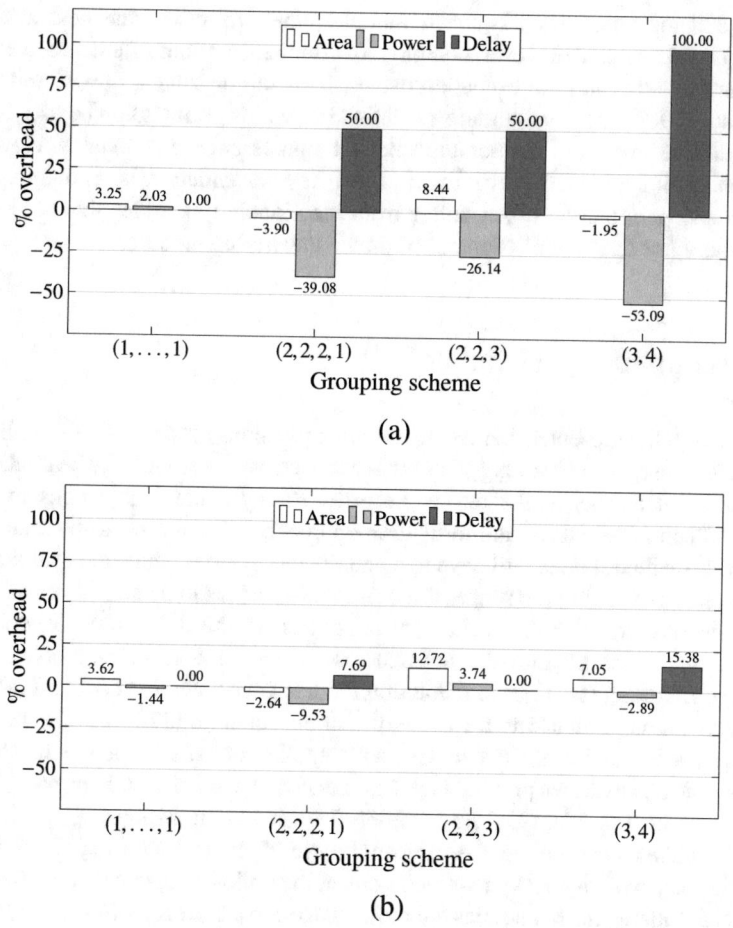

Fig. 8.13 Control-Lock evaluation at high-performance points for the RI5CY (**a**) Decoder and (**b**) ALU

deploy more complex grouping schemes as these yield a relatively low-cost impact on the overall design while providing the highest level of security.

8.2.6   Related Work

To the best of our knowledge, Control-Lock is the first technique to protect against hardware Trojans that exploit critical inter-module control signals. However, alternative approaches can be found in the existing encryption solutions. For example, selected signals can be protected using a fully fledged (data) encryption scheme, such as the Advanced Encryption Standard (AES) cipher [45]. In terms of protecting

the data being transmitted between modules, the AES offers the best available
security level thus far. However, common AES implementations often exceed 20,000
gates and require multiple execution cycles, inducing high area, power, and delay
overheads [190]. Another alternative solution lies within permutation boxes (p-box).
These can permute and reassemble selected signals between modules. However,
p-boxes exhibit constant behavior, i.e., no key dependency is available. Both
ciphers and permutation boxes suffer from identifiable typologies that are neither
structurally nor functionally dependent on the underlying modules.

8.3   Limitations and Outlook

Even though the presented framework and its extensions have proven its applicability in an industry-ready setting, its security aspects will undoubtedly be challenged
against novel attacks in the future. Security, in any context, is a constant arms
race between novel attack and mitigation approaches. Therefore, to be secure, the
presented methodologies will have to stand the test of time. Thus, the evaluation of
potential threats within the scope of the framework needs to be actively continued.

In the context of Inter-Lock, further studies of the ILCs have to verify the
removal resilience beyond the activities described in Sect. 8.1.4. For example,
Machine Learning (ML)-based prediction models have recently shown a high level
of effectiveness in attacking a variety of locking schemes [170]. Similar concepts
based on ML can be drafted to try isolating ILC outputs from the rest of the
design. The observations provided by these attack mechanisms could provide further
insights on how to adapt and improve the ILC generation. Moreover, specific per-
module schemes can be steered to enable more efficient design implementations.
For example, avoiding heavy manipulations of the critical path can hamper the delay
overhead. Finally, further studies are required to gain a correct measure of a suitable
number of internally generated keys per source module depending on its structure
and functionality in order to yield a more obfuscated design.

The Control-Lock methodology can potentially induce high costs depending
on the number of selected signals. To stay within acceptable boundaries, it is
necessary to limit the application to high-priority inter-module signals. However,
a straightforward identification procedure has not been developed yet. Thus, an
interesting research opportunity lies within the assessment and prioritization of
in-processor signals in the context of HT prevention. Furthermore, similar to Inter-
Lock, the structural aspects of the CLC integration must be further evaluated for
potential removal attacks.

8.4 Synopsis

The first part of this chapter focused on the Inter-Lock methodology. This extension addresses the challenges of protecting complex, multi-module HW designs against reverse engineering. The locking policy exploits inter-module dependencies to thwart attacks on separate modules, thereby shifting the main attack vector to the overall design. The cost implications of the methodology have been evaluated on a range of key lengths for a concrete technology node. Furthermore, the application of Inter-Lock and the entire locking framework has been demonstrated in an industrial setting through the production of the MiG-V processor core.

The second part of the chapter introduced Control-Lock, a methodology for preventing software-controlled hardware Trojans that exploit critical control signals in processor designs. Control-Lock encrypts and decrypts a set of inter-module signals based on the proposed randomized key-dependent netlist generation flow. The cost analysis was performed on a RISC-V processor case study, assuming a denial of service HT. Both methodologies encourage further development and research of effective protection schemes for complex processor designs, thereby focusing on tangible attack scenarios.

Part IV
Machine Learning for Logic Locking

Chapter 9
Security Evaluation with Machine Learning

The proliferation of efficient, easy-to-use, and easy-to-access Machine Learning (ML) models has started to have a profound impact on the domain of logic locking. Researchers have realized that many security characteristics of contemporary logic locking schemes are wrongfully assumed to be true, specifically due to successful attacks by ML-driven deobfuscation techniques. The reason that these assumptions were not verified before is that a successful verification often requires detecting patterns in complex, large data sets—a task that is difficult for humans. Moreover, processing Hardware (HW) designs at different levels with ML is still an evolving field of research. Thus, only recent developments have shed light on the application of ML in logic locking. In the context of security evaluations, the latest efforts have been targeted toward the removal of the locking circuitry and the retrieval of the correct activation key through the application of ML. To understand how ML can be adapted to evaluate the security features of logic locking, in this chapter, we introduce the following:

- The introduction of SnapShot, an ML-based Oracle-Less (OL) attack on logic locking that utilizes Artificial neural networks (ANNs) to predict key values [169, 170]

The rest of this chapter is organized as follows. Section 9.1 introduces the blueprint of an ML-driven attack on logic locking. Further implementation details for each component of the attack are discussed in Sect. 9.2. The attack is evaluated in Sect. 9.3. Limitations are discussed in Sect. 9.4.

9.1 Constructing an ML-Driven Attack

As discussed in Sect. 4.7, OL attacks are currently still representing the stronger attack model, requiring an attacker to unlock and reverse engineer a target design

© The Author(s), under exclusive license to Springer Nature Switzerland AG 2023
D. Sisejkovic, R. Leupers, *Logic Locking*,
https://doi.org/10.1007/978-3-031-19123-7_9

only based on the logic-locked design instance. Thus, an attack must be able to learn a correlation between *the locking-induced structural residue* and *the correct activation key*. This specific task is well suited for deep learning techniques, such as convolutional neural networks, that can automatically extract features and learn a correlation even beyond the capabilities of a human observer.

To explore the potential of ML in compromising logic locking, we introduce the *SnapShot* attack. SnapShot is the first of its kind to directly predict a key value by analyzing "snapshots," i.e., excerpts of a locked gate-level netlist, thereby utilizing neuroevolutionary and deep learning methods. The significance of SnapShot lies in the following:

- The attack operates in the OL regime; hence, an activated oracle is not required.
- The attack is applicable for any known gate-insertion-based scheme.
- The attack supports both combinational and sequential circuits.
- The attack has a linear time complexity with respect to the key length.
- The attack can be customized to support a range of supervised learning models.
- Finally, the attack can be adjusted to a range of HW design levels.

Note that the attack flow introduced in this chapter can be used as a blueprint for ML-based attack construction. Preliminaries on deep learning, neuroevolution, and relevant neural network architectures can be found in Appendixs C.1 and C.2.

9.2   Attack Flow

The objective of SnapShot is the immediate prediction of a key-bit value based on the extracted *locality* around each key gate. A locality represents a key-dependent portion of the netlist. The reasoning for this attack is as follows. As per the attack assumptions described in Sect. 4.7, the attacker is aware of the locations of all key inputs. For each key input, the exact key gate (or netlist structure) can be identified by following the key-input signal. This can be repeated for each bit of the key individually. Since current gate-insertion-based locking schemes only induce limited local changes to the netlist [36], it is possible for the underlying ML model to predict which key-bit value to expect for a given locality that is associated with a specific key input.

The SnapShot attack flow is presented in Fig. 9.1. The input consists of the number of samples for data generation (N), the key length (K), and the target netlist. The output is the predicted activation key for the target netlist. The following introduces further details on each step.

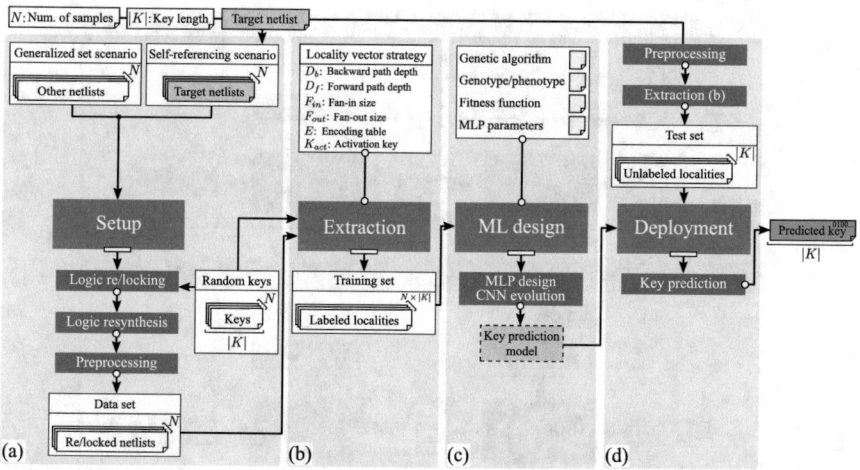

Fig. 9.1 SnapShot attack flow: (**a**) attack setup, (**b**) localities extraction, (**c**) ML model design, and (**d**) deployment

9.2.1 Setup: What Is the Attack Scenario?

The setup stage of SnapShot prepares the *data set* upon which a *training set* is extracted in the following stage (Fig. 9.1a). The data set comprises a set of locked post-resynthesis benchmark netlists. The training set represents the labeled localities in the form of vectors that are extracted from the data set for each key input. This setup can result in two realistic attack scenarios as follows.

Generalized Set Scenario (GSS) In GSS, the ML model in SnapShot is trained based on the extracted data from a generalized data set that is assembled by locking a large number of netlists (potentially different from target) with random keys (Fig. 9.2a). The reasoning behind GSS is that, in a real-life attack, any available circuits (open-source or in-house designs) can be utilized for training. Note that the set of netlists can also include already locked designs, even with unknown keys.

Self-Referencing Scenario (SRS) In SRS, the attacker can create a data set by *relocking*, i.e., self-referencing, the already locked target with additional keys (Fig. 9.2b). Hence, the resulting data set contains relocked copies of the target with additional key inputs. These additional key inputs with known key values are used to train the ML model to predict the key of the target netlist.

After the data set is generated, all sample netlists are preprocessed and transformed into a generic, technology-independent format. This process is detailed in Sect. 7.3.

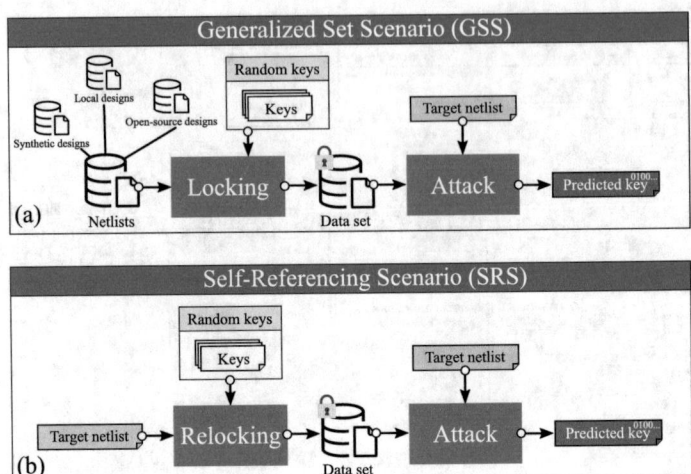

Fig. 9.2 ML-based attack scenarios

9.2.2 Extraction: What to Present to the ML Model?

The extraction stage involves the generation of the training set (Fig. 9.1b). The training set contains the labeled localities that are extracted from the data set. The localities are a compressed representation of the netlist subgraph that is functionally affected by a key input. The extraction and representation format of localities can be implemented in various ways. However, the locality must contain all the relevant information that could potentially be related to the key value, such as the number of gates, the gate types, and their relationship. To exemplify the extraction process, in the following, we introduce the extraction mechanics for two different scheme types: XOR/XNOR-based and MUX-based locking. For both scheme types, the procedure results in a set of *locality vectors* that offer a suitable representation in the form of unstructured image data, which is specifically advantageous for neural networks.

9.2.2.1 Locality Vector Extraction

As previously discussed, neural networks are well suited for operating on unstructured image data. Hence, it is advantageous to represent localities in a form that can be translated into images. Hereby, the prediction problem is modeled as the classification of the extracted image as key bit 0 or 1. For this purpose, we introduce the concept of a locality vector that represents the extracted subgraph as a 1-dimensional sequence of integers. The representation works as follows. Each entry represents a single location in the netlist. A location can be filled by a gate or remain empty. The value of the entry simultaneously determines the gate type and

Table 9.1 Locality vector encoding table

Gate type	NOT	AND	NAND	OR	XOR	NOR	XNOR	BUF	FF	MUX
Code	1	2	3	4	5	6	7	8	9	10

Algorithm 9.1: LVE: Locality Vector Extraction

Input: Locked generic netlist (Net), backward path depth (D_b), forward path depth (D_f), fan-in (F_{in}), fan-out (F_{out}), encoding table (E), and activation key (optional) (K_{act})

Output: Set of locality vectors (L)

```
1  L ← [∅]                                    // Prepare locality vector set
2  K ← |Net.K_in|                             // Length of key input
3  T ← {Backward, Forward}                    // BFS type
4
5  for i = 0 to K do
6      kg_i ← GATECONNECTEDTO(Net.K_in[i])
7      g_i ← INPUTGATESOF(kg_i)[1]
8
       /* Extract values                                              */
9      l_b ← BFS(T.Backward, g_i, E, D_b, F_in)
10     l_kg ← E[TYPE(kg_i)]
11     l_f ← BFS(T.Forward, kg_i, E, D_f, F_out)
12
       /* Merge values                                                */
13     L[i] ← {l_b, l_kg, l_f}
14
       /* If K_act is given: label the locality                       */
15     if K_act != {∅} then
16         L[i] ← {K_act[i], L[i]}
17     end
18 end
19 return L
```

---

its existence in that location. These properties are uniquely identified by an encoding of the gate types. In this book, we selected the encoding in Table 9.1. Note that the existence of a gate is encoded indirectly; an empty location is denoted with the value 0; otherwise, the location is filled by a gate of a particular type.

The Locality Vector Extraction (LVE) procedure is presented in Algorithm 9.1. The input consists of the locked netlist, backward and forward path depth, maximum fan-in and fan-out size, encoding table, and activation key. The output is the set of extracted locality vectors. The main LVE loop repeats for every 1-bit key input (line 5). First, LVE locates the first key gate (kg_i) connected to the currently observed key input ($Net.K_{in}[i]$) as well as the first non-key input (g_i) connected to kg_i (lines 6 and 7). Second, LVE extracts three sections of a locality vector as follows. The first section is derived through a Breadth-First Search (BFS) starting from g_i toward the primary inputs (backward path, line 9). The second section extracts the encoding of the central gate kg_i (line 10). The third section is obtained by performing a BFS starting from kg_i in the direction of the primary outputs (forward path, line 11).

Algorithm 9.2: BFS Extraction Procedure

Input: BFS type (T_{bfs}), root (R), encoding table (E), max depth (D), and fan-in/out (F)
Output: Locality vector (l)

```
 1  l ← {∅}; Q ← [∅]              // Prepare locality vector and FIFO queue
 2  done ← FALSE; D_i ← 1         // Prepare control variable and current
    depth
 3  Q.ENQUEUE(R)                                              // Add root gate
 4  R.Visited← TRUE                                     // Set root as visited
 5
 6  while !ISEMPTY(Q) and !done do
 7  │   g_i ← Q.DEQUEUE( )
 8  │   N ← [∅]
 9  │
    │   /* Returns {∅} for empty gate & primary IOs              */
10  │   if T_bfs == T.Backward then
11  │   │   N ← INPUTGATESOF(g_i)
12  │   else
13  │   │   N ← OUTPUTGATESOF(g_i)
14  │   end
15  │
    │   /* Ensures: gate has F ins or outs                       */
16  │   if j = |N| to F then
17  │   │   N[j] ← NEWEMPTYGATE( )
18  │   end
19  │
    │   /* Visit neighbors                                       */
20  │   for j = 0 to F do
21  │   │   N_j ← N[j]
22  │   │   if !N_j.Visited then
23  │   │   │   Q.ENQUEUE(N_j)
24  │   │   │   N_j.Visited ← TRUE
25  │   │   │
        │   │   /* Extract value                                 */
26  │   │   │   e ← E[TYPE(N_j)]
27  │   │   │   if T_bfs == T.Backward then
28  │   │   │   │   l.PREPEND(E[e])
29  │   │   │   else
30  │   │   │   │   l.APPEND(E[e])
31  │   │   │   end
32  │   │   end
33  │   end
34  │
    │   /* Check if max depth reached                            */
35  │   D_i ← D_i + 1
36  │   if D_i == D then
37  │   │   done ← TRUE
38  │   end
39  end
40  return l
```

Finally, all three sections are concatenated to form a locality vector (line 13). In case the activation key is given, all extracted localities are labeled with their respective correct key bit (lines 15–17). The activation key is an optional input argument since the extraction procedure can be utilized to generate a training set (labeled vectors) as well as a test set (unlabeled vectors).

The BFS-based extraction policy used in LVE is presented in Algorithm 9.2. BFS receives the following inputs: the BFS type (T_{bfs}), the root node (R), the selected encoding table (E), the maximum allowed search depth for the backward and forward path, and the fan-in and fan-out size per gate. The depths and fan sizes are represented by the variables D and F, respectively. The traversal is based on a standard BFS implementation with variable search properties, as discussed in the following. The search direction is directed by T_{bfs}. In a backward search, the traversal follows all input gates of the current gate g_i (line 11). In a forward search, the traversal follows all output gates of g_i (line 13). The algorithm assumes a fixed number of possible inputs and outputs per gate, as defined by F. As the preprocessing step in the setup stage of SnapShot ensures that each gate has a maximum of two inputs (see Sect. 9.2.1), $F = 2$ for a backward search. However, a single gate can have any number of outputs. Thus, the number of output gates has to be manually selected. Fixing the number of inputs and outputs ensures that each locality has the *same length* regardless of the subgraph it represents. In cases when a gate deviates from the selected fixed values, the BFS algorithm inserts empty gates (lines 16–18) encoded with 0. BFS terminates once the selected search depth is reached (lines 35–38). The locality vector l is constructed by storing the encoding for each visited neighbor (lines 26–31). The neighbor encoding is either prepended (for backward BFS) or appended (for forward BFS) to l to reflect the topological order of the gates. Hence, looking from the central key gate, the gates that are closer to the primary inputs are placed toward the left end of l, and the gates that are closer to the primary outputs are placed toward the right end of l. More details are available in [169].

9.2.2.2 Localities Extraction for XOR/XNOR-Based Locking

An example of the locality extraction for a 1-bit key input for XOR/XNOR-based logic locking is presented in Fig. 9.3. The extraction parameters are set to $D_b = 2$, $D_f = 3$, $F_{in} = 2$, and $F_{out} = 3$. Based on the encoding in Table 9.1, the discussed extraction policy extracts one labeled locality vector in the format $[K_{act}[i], l_b, l_{kg}, l_f]$. The example also exemplifies the insertion of empty gates (marked with 0).

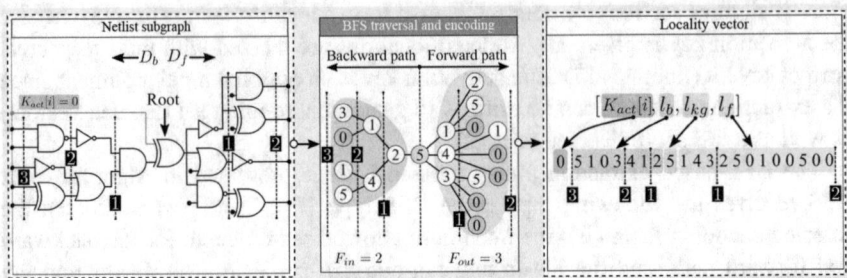

Fig. 9.3 Example of localities extraction for XOR/XNOR-based locking

9.2.3   ML Design: Which Model to Select?

The next stage of the SnapShot attack flow is the ML design (Fig. 9.1c). This design stage includes the *modeling* and *training* of a selected ML model *based on the given training set*, which has been extracted in the previous stage. The output of the ML design stage is a fully trained key prediction model.

Note that the SnapShot attack is, in principle, independent of the specifics of ML models, hence allowing for model exploration. However, the nature of the extracted locality vectors suggests that artificial neural networks, specifically Convolutional Neural Networks (CNNs), offer a suitable apparatus for this prediction problem. A motivational analysis is provided in Sect. 9.2.3.1.

Nevertheless, any ML model that fits the task can be used within SnapShot. For example, recent efforts have demonstrated the applicability of SnapShot to Register-Transfer Level (RTL)-based locking by utilizing automatic ML (auto-ML) libraries [165].

9.2.3.1   Locality Vectors as Image Data

As mentioned in Sect. 9.2.3, CNNs are a suitable tool to predict key values based on locality vectors. To justify this claim, in the following, we present a motivational example. First, a randomly selected circuit is logic-locked with the Random Logic Locking (RLL) scheme using 64-bit keys. Second, 400 localities for both labels 0 and 1 are extracted. Finally, all 0 and 1 localities are stacked horizontally to form two images, as shown in Fig. 9.4: the first one represents all 0-labeled vectors, and the second represents all 1-labeled vectors. Through careful examination, it is possible to detect repeating patterns that are *specific to each key-bit value*. Example patterns are emphasized in both images. Here, the correlation between the visualized patterns and the key-bit values is easily recognizable even for a human. Since extracting features and learning a correlation in data are core features of CNNs, it stands to reason that CNNs are a suitable tool for this prediction problem.

Fig. 9.4 Locality vectors in
the form of images for 0 and
1 key-bit values. (**a**) 0-labeled
localities. (**b**) 1-Labeled
localities

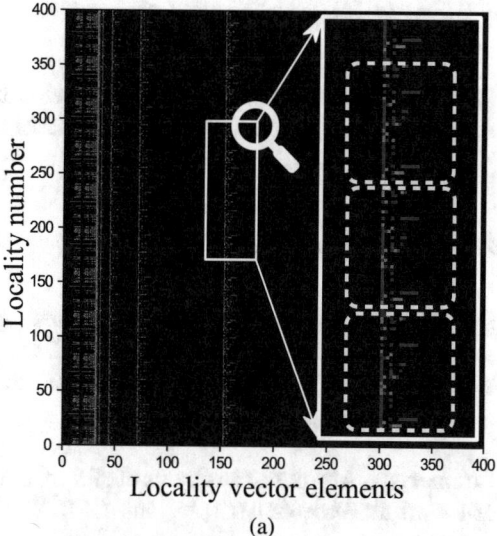

(a)

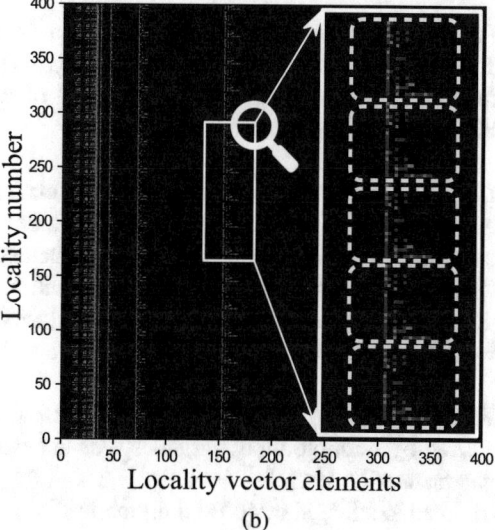

(b)

9.2.4   Deployment: How to Execute the Attack?

The final stage of SnapShot is the deployment of the attack (Fig. 9.1d). The input to
this stage is the target test set and the trained prediction model. The test set consists
of *unlabeled* locality vectors, which have been extracted from the target netlist. The
output of the deployed attack is a set of predicted key bits. Since each key-bit value
can be attacked, i.e., predicted individually, SnapShot is easily scalable to large
designs and a high number of key inputs.

9.3   Evaluation

The following describes the evaluation setup that was deployed to evaluate the efficacy of SnapShot on a range of benchmarks.

9.3.1   *Experimental Environment*

Benchmarks To evaluate SnapShot, three open-source benchmark sets were selected (Table 9.2): ISCAS'85 [31], ITC'99 [44], and a set of modules from the 64-bit RISC-V Ariane core [225]. Both the generalized set scenario and the self-referencing scenario were evaluated.

Comparison Attacks The results of SRS are relevant for comparison with the state-of-the-art ML-based SAIL [36] and OMLA [13] attacks. See Sect. 5.1 for more details.

Environment The attack implementation is constructed on the Python-based TensorFlow framework for deep learning [1]. The evaluation was performed on an Intel Core i5 CPU@3.2 GHz with 16 GB of RAM and a single Nvidia GeForce RTX 2080 Ti graphic card.

Locking Scheme RLL was selected for evaluating SnapShot. The reasoning is twofold. First, RLL represents a *superset* of other XOR/XNOR-based policies, hence covering a broad spectrum of locking schemes. Second, the security concept of XOR/XNOR locking is often embedded within a wide range of modern schemes, thus allowing the evaluation to draw conclusions about more complex policies as well. An overview of locking policies is available in Chap. 5.

Extraction Setup As detailed in Sect. 9.2.2.1, the extraction methodology can be steered by multiple input variables. For the evaluation, we selected the following parameters: $D_b = 5$, $D_f = 5$, $F_{in} = 2$, and $F_{out} = 3$. The extracted localities were trimmed to a length of 400 and normalized to values between 0 and 1.

Performance Measure The success of SnapShot is evaluated in terms of the Key Prediction Accuracy (KPA). This measure equals the percentage of correctly predicted key bits for a target netlist, where each key-bit input is predicted individually. For example, a KPA of 80% indicates that 80% of the bits of a single key are *correctly predicted*. Note that this measure does not indicate *which* bits are correctly predicted.

9.3.2 Model Setup

As mentioned earlier, SnapShot is independent of the underlying ML model. However, due to the nature of the extracted data, as discussed in Sect. 9.2.3, we explored two models: Multi-Layer Perceptrons (MLPs) and CNNs.

MLPs are shallow neural networks with only a few hidden layers. Due to their simplicity, we manually adjusted the MLP architecture by varying the number of hidden layers and the number of units (neurons) per layer for both SRS and GSS. After the exploration, we fixed one architecture for each attack scenario based on the best performance achieved in the tuning phase.

CNNs are more complex neural networks that incorporate an automatic feature extraction based on which the classification is done. Their complexity, however, comes with a disadvantage; a large number of hyperparameters require tuning. To mitigate this challenge, we utilize a neuroevolutionary approach to automatically evolve CNN architectures that are suitable for the given prediction problem. The process of architecture evolution is presented in Appendix C.3. More details on the exact parameters for the utilized MLPs and CNNs are available in Appendix D.4.

9.3.3 Data Preparation

As discussed earlier, the evaluation focuses on two attack scenarios: GSS and SRS. For both scenarios, a sufficient amount of data samples must be generated to deploy the SnapShot attack. For GSS, the training set was compiled of all mentioned ISCAS'85 and ITC'99 benchmarks from Table 9.2. Each training benchmark was locked 1000 times with randomly generated 64-bit keys, resulting in a training set of 832,000 *labeled* training localities. The test set was assembled based on the Ariane modules, each locked 1000 times with randomly generated 64-bit keys, hence resulting in 448,000 *unlabeled* test localities (64,000 per target netlist).

Table 9.2 Benchmark circuits

(a) ISCAS'85

IC	#Ins	#Gates	#Outs
c1355	41	546	32
c1908	33	880	25
c2670	233	1193	140
c3540	50	1669	22
c5315	178	2307	123
c6288	32	2416	32
c7552	270	3512	108

(b) ITC'99

IC	#Ins	#Gates	#FFs	#Outs
b15	37	6931	447	70
b21	34	7931	494	22
b22	34	12128	709	22
b17	38	21191	1407	97
b18	37	49293	3308	30
b19	47	98726	6618	40

(c) Ariane RISC-V

IC	#Ins	#Gates	#Outs
iscan	32	240	139
commit	985	1584	417
brunit	342	1655	328
decoder	518	2169	362
pcselect	521	3333	128
brpredict	814	4669	333
alu	206	7412	65

In SRS, each target netlist is self-referenced to generate the training set. The test set was generated by locking all ISCAS'85 and ITC'99 benchmarks with 64-bit keys. The training set was generated by *relocking* the locked targets repeatedly 1000 times with additional 64-bit keys. Hence, the training set consisted of 64,000 training localities per benchmark and iteration instance.

9.3.4 Results: Generalized Set Scenario

As presented in Fig. 9.6a, the average test KPAs for GSS across all modules are 57.71% for MLP and 61.56% for CNN. The results indicate the feasibility of learning from a set of locked netlists to predict the key of a previously unseen target with an average increase in accuracy of up to 11.56 percentage points compared to a random guess. Moreover, the CNN model offers a relatively small performance gain of 3.85 percentage points compared to MLP. Even though the KPA values are lower for GSS compared to SRS as discussed in the following section, GSS offers one advantage; only one universal training set is required to attack any locked target. Hence, the training can, in principle, be performed only once. Note that KPA values for SAIL and OMLA were not available in the literature for the GSS case. The KPA values for each benchmark for GSS are presented in Fig. 9.5a.

9.3.5 Results: Self-Referencing Scenario

To compare SnapShot against the state-of-the-art SAIL and OMLA attacks in SRS, we present the average test KPA comparison in Fig. 9.6b. This comparison was performed on all mentioned ISCAS'85 benchmarks.[1]

9.3.5.1   SnapShot vs. SAIL

The average CNN-SnapShot KPA is 82.60%, yielding a significant improvement of 10.49 percentage points compared to SAIL. Moreover, as presented in Fig. 9.5b, c, the SnapShot attack outperforms SAIL *on all evaluated benchmarks*. Furthermore, CNN-SnapShot achieves a minimum KPA of 78.43% (for c2670), whereas SAIL achieves a KPA of 55.83% (for c1355), only 5.83 percentage points better than a random guess. This suggests that SnapShot can achieve high KPA values regardless of the circuit type. The MLP-based SnapShot attack results in a relatively low average KPA of 68.00%, suggesting that the more expressive CNN is the more suitable ML model for this task.

---

[1] Results of the SAIL attack on other benchmarks were not available in the literature.

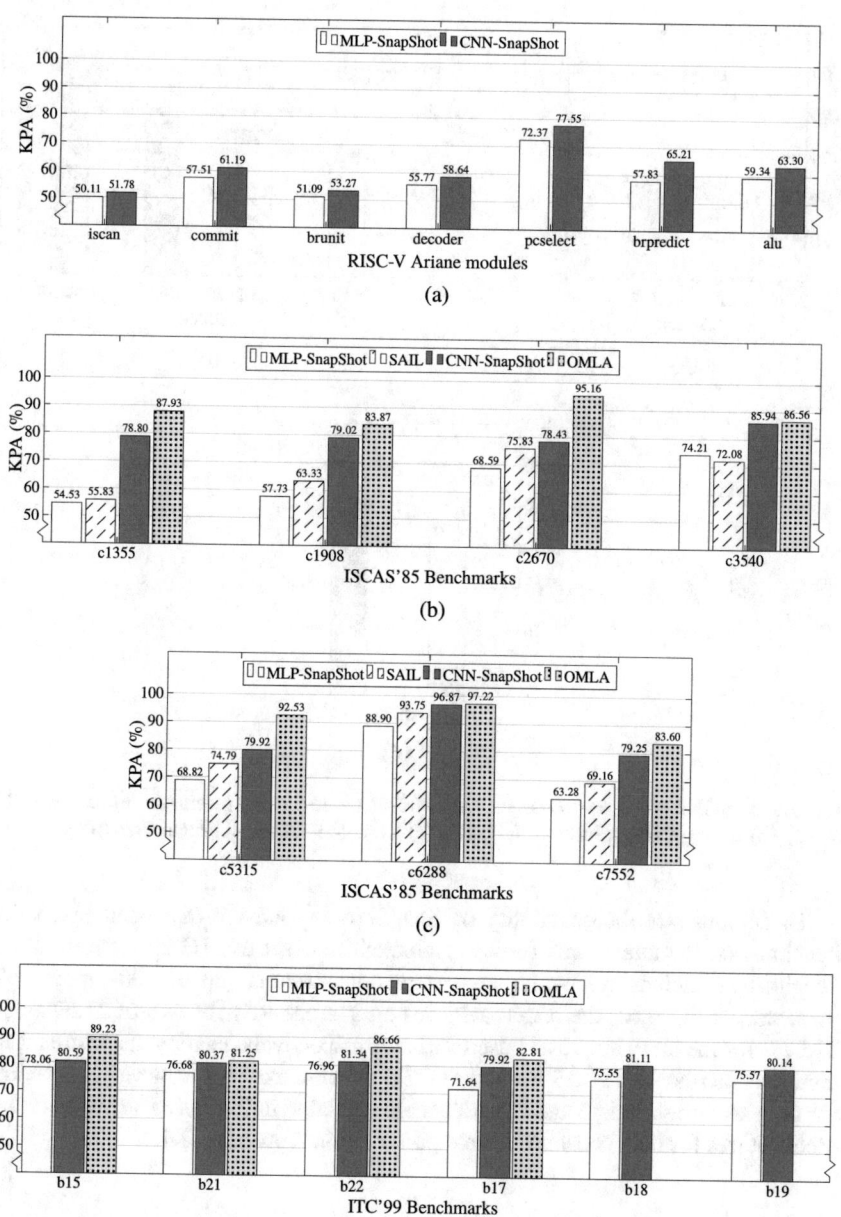

Fig. 9.5 SnapShot attack evaluation on XOR/XNOR-based locking (per benchmark). (**a**) GSS evaluation. (**b**) SRS evaluation (part 1) (**c**) SRS evaluation (part 2). (**d**) SRS evaluation (part 3)

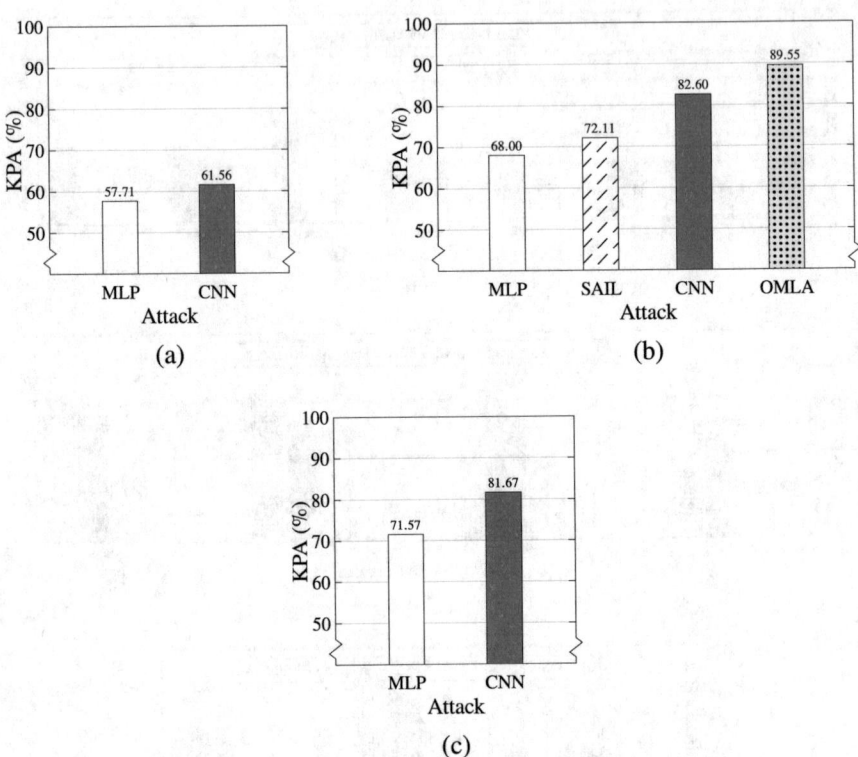

Fig. 9.6 SnapShot attack evaluation on XOR/XNOR-based locking (average). (**a**) Average KPA (GSS). (**b**) Average KPA for ISCAS'85 (SRS) (**c**) Average KPA for all benchmarks (SRS)

To demonstrate the scalability of SnapShot beyond the discussed ISCAS'85 benchmarks, the attack was further evaluated on a set of ITC'99 circuits. These benchmarks include relatively large sequential circuits (up to 100k gates). As presented in Fig. 9.6c, the attack achieves an average KPA for SRS of 71.57% and 81.67% for the MLP- and CNN-based variants, respectively. Hereby, the results were computed across all ISCAS'85 and ITC'99 benchmarks. All achieved KPA values for both scenarios and all attack variants are available in Fig. 9.5. The evolved CNN architectures for both GSS and SRS are presented in Appendix D.4.3.

9.3.5.2   SnapShot vs. OMLA

OMLA outperforms the other attacks across all evaluated benchmarks reaching an average KPA of 89.55%, yielding an improvement of 6.95 percentage points over SnapShot and 17.44 percentage points over SAIL (Fig. 9.6b). Note that these KPA values do not account for the evaluation results over the ITC'99 benchmarks

(Fig. 9.5d), as OMLA-generated KPA values over sequential benchmark were not available in the literature.

9.3.6 Attack Comparison

Apart from the higher attack accuracy, SnapShot has multiple advantages over SAIL. First, SAIL is exclusively applicable to XOR/XNOR-based locking since the attack relies on the inherent security properties of this locking policy. In comparison, the extraction stage of SnapShot can be adjusted to capture a wide range of locking-induced changes in the netlist. Second, SAIL learns to reconstruct the synthesis changes to predict the original netlist structures. Hence, its success heavily depends on the complexity and features of the deployed synthesis tool. In comparison, SnapShot allows a direct key prediction by operating on the post-resynthesis netlists, thus offering a simpler and more flexible attack vector.

The results indicate that OMLA is a more efficient attack compared to SnapShot and SAIL. This is caused by OMLA's underlying ML model—Graph Neural Networks (GNNs). The operation mode of GNNs relies on a graph representation of the input data, hence offering a more suitable ML model for processing gate-level netlists compared to custom-made locality vectors. Note, however, that *the overall attack flow of SnapShot is not bound to CNNs and locality vectors*. In fact, SnapShot represents an attack blueprint, where its internal extraction and ML design steps can be adjusted to fit the HW design level and the scheme type. For example, in OMLA, the extraction is represented by gathering labeled subgraphs, whereas the ML design incorporates a construction of a graph neural network.

9.4 Limitations and Outlook

As presented in this chapter, OL ML-based attacks can potentially yield a high prediction accuracy. However, one limitation remains: the *uncertainty* of the prediction for each key bit. Nevertheless, the output of these attacks can be used as a seed for subsequent attacks to further refine the key as well as facilitate the process of identifying components within the process of reverse engineering. In terms of SnapShot, its attack flow offers multiple improvement points. First, its ML model can be extended to include GNNs, which offers a suitable foundation to analyze graph structures in the context of logic locking as showcased in [10, 13]. Second, the extraction procedure as well as the locality representation can be further adjusted to capture more key-related data with respect to specific locking policies, potentially leading to higher accuracy values. Finally, GSS outlines a new research opportunity—assembling a generic training set that covers a large range of locked benchmarks. In the long run, this set can be used to automatically design resilient locking schemes.

9.5   Synopsis

This chapter introduced a blueprint for an ML-based, oracle-less attack on logic locking. The attack is the first of its kind to immediately predict a key value out of a locked netlist using ML models. The attack concept exploits the key-related structural changes to extract a correlation between the scheme-induced netlist change and the correct key value. To facilitate the attack design, we further introduced a neuroevolutionary process for the automatic construction of suitable neural networks for the selected prediction task. The evaluation of SnapShot has uncovered an important structural vulnerability of XOR/XNOR-based locking schemes, thereby initiating the concept of learning resilience within the design of logic locking. The comparison of SnapShot with SAIL and OMLA confirmed that GNNs offer a more suitable tool to process and attack locked gate-level netlists compared to MLPs and CNNs. In the following chapter, we will focus on the problem of modeling and testing learning resilience of locking schemes as well as the construction of learning-resilient policies.

Chapter 10
Designing Deceptive Logic Locking

Even though multiple Machine Learning (ML)-based attacks have been introduced in recent years, the theoretical concepts of ML-exploitable leakage in logic locking have not been established. Consequently, a gap in the design of locking policies that are resilient against ML-based attacks has emerged. To advance the state of logic locking in the ML era, in this chapter, we introduce the following [166]:

- The first theoretical concept for evaluating learning resilience in logic locking.
- An Oracle-Less (OL) structural attack on Multiplexer (MUX)-based locking.
- The process of constructing a MUX-based locking policy that is resilient against the SnapShot attack. This process exemplifies the application of the theoretical concepts for the construction and evaluation of schemes in the ML era.

The application of machine learning techniques has started to have a deep impact on logic locking. In addition to SnapShot, a handful of recent works have successfully deployed ML models to evaluate the security of logic locking [10, 13, 36, 40, 88, 165, 170]. Nevertheless, so far, there has been a major gap in the theoretical means of uncovering the source of ML-exploitable leakage as well as in designing logic locking that is resilient against ML-based attacks. To address this challenge, in the following, we introduce the first theoretical concept to test learning resilience and apply the lessons learned for the design of the first empirically evaluated ML-resilient locking policy. Hereby, the term *learning resilience* is defined as follows:

Definition 10.1 **Learning resilience** defines the resilience of logic locking against learning-based attacks that capture *locking-induced structural residue* that can lead to *key-related information leakage*.

© The Author(s), under exclusive license to Springer Nature Switzerland AG 2023
D. Sisejkovic, R. Leupers, *Logic Locking*,
https://doi.org/10.1007/978-3-031-19123-7_10

The reasoning for this definition comes from the nature of ML-based attacks that typically operate in an oracle-less mode, hence focusing on structural netlist features. Note that if a scheme is learning-resilient, we refer to it as *deceptive*.[1]

10.1 The Learning-Resilience Test

To evaluate logic locking in the context of learning resilience, we devise a test that analyzes what structural changes are induced by a scheme, thereby considering *two netlist variants*. The first variant considers netlists that only contain *a single gate type*. The second variant includes netlists that consist of a randomly selected and well-distributed amount of *all gate types*. Therefore, these two variants represent two ends of the spectrum of netlist structures: *regular* and *irregular*. Hereby, the regularity describes the repetition of equivalent logic structures throughout the netlist. The rationale for analyzing these structural extremes is as follows. In practice, netlists can be placed between the two variants. Thus, a learning-resilient scheme must ensure resilient properties regardless of the underlying structural traits of the target netlist. However, testing for all possible netlist variants is not a viable procedure. Therefore, by looking at the structural extremes, we can emphasize potential leakage points that otherwise would have been missed due to the specifics of the target netlist.[2]

Using the two structural extremes, we can devise two tests for learning resilience. The regular netlists are represented by the AND Netlist Test (ANT) and the irregular by the Random Netlist Test (RNT). ANT and RNT offer a means of uncovering potential structure-related leakage at the two spectrum ends. Both ANT and RNT are implemented in the form of a learning game. Thereby, both tests follow the same testing procedure but operate on different netlist structures. The learning game includes the interaction of two players: Trusted (**T**) and Untrusted (**U**). Through an iterative process, the goal of **U** is to learn to predict the correct key by analyzing the locked netlist provided by **T**. *If a locking scheme fails at least one test, i.e., **U** is able to make an educated guess about the key,*[3] *the locking scheme is evaluated as* **conclusively vulnerable**, *otherwise it* **might be learning-resilient**. Hence, the test can only make a conclusive decision when a scheme is *not* resilient. One iteration of the game is visualized in Fig. 10.1. The game consists of the following steps:

1. **T** randomly generates a netlist (*Net*) that is either regular (for ANT) or irregular (for RNT), where each node can have any number of input or output connections.

---

[1] The term "deceptiveness" must not be confused with the functional output deceptiveness for incorrect keys (discussed in Chap. 6).

[2] Note that most hardware designs are more likely structurally regular. Examples include adder, S-box, multiplier, decoder, machine learning, and other implementations. The success of ML-based, structural attacks supports this claim. In contrast, irregular structures are mostly found in the form of control logic.

[3] An educated guess is possible if the guessing accuracy is higher than 50%.

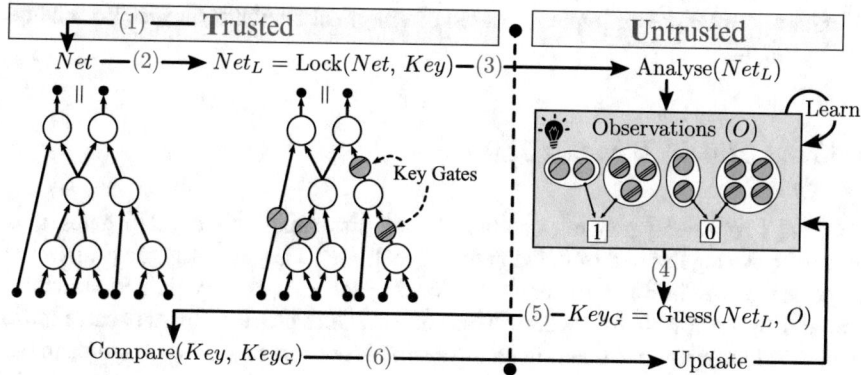

Fig. 10.1 One iteration of the learning-resilience test

2. **T** locks the netlist (Net_L) with a selected scheme using a randomly selected key (Key).
3. **T** sends the locked netlist to **U** for analysis.
4. **U** guesses the key based on the collected observations.
5. **U** sends the guessed key (Key_G) to **T** for comparison.
6. **T** responds whether the guess is correct or not for each key-bit value individually. Note that **T** does not provide the reason why a bit is correctly or falsely guessed.

At the end of each iteration, **U** updates his guessing capabilities with new observations. Therefore, if the scheme under attack leaks information, **U** might be able to make educated guesses after enough observations were collected. If the guessing accuracy remains 50%, **U** is not able to recognize any evident leakage points.

Furthermore, the game operates under the following assumptions: (1) **T** always uses the same locking policy and (2) the original netlist is different in each iteration. (1) ensures that **U** is able to learn if a leakage exists, otherwise **T** could always deploy a different scheme. (2) prevents learning from different key values for identical netlist locations. The game terminates when **U** achieves the desired success rate (guessing accuracy).

Moreover, the game assumes that **U** has some knowledge about the initial netlist structure. This is evidently not in line with an actual attack scenario since an attacker should not be informed about the design details. However, this assumption has a favorable connotation; it *amplifies* the locking-induced changes. *Therefore, any conclusions made about a scheme in ANT or RNT hold also in the general, for the attacker less invasive case.*

Testing whether a human adversary (or, ultimately, a machine learning system) is able to challenge logic locking through learning-based attacks is a fundamental step in the design process since it tests the basic security foundation: *can the adversary guess the key based on the observed change?* Thus, the introduced theoretical tests offer a simple yet powerful approach to uncover elemental security flaws in logic locking; even before an implementation or empirical results for a specific scheme are available.

10.1.1   The AND Netlist Test

In ANT, **T** repeatedly generates netlists that consist exclusively of AND gates, thus covering the regular end of the structural spectrum.[4] The regularity maximizes the exposure of the intrinsic features of a locking scheme, as only limited structural variability is available for locking. This netlist feature provided by ANT can expose potential leakage points that rely on the availability of specific gate structures in the design.

10.1.2   The Random Netlist Test

In RNT, **T** repeatedly generates netlists that consist of a variety of randomly selected gate types, hence representing the irregular end of the structural spectrum. In RNT, a particular locking policy might not leak, i.e., reveal information about the key due to the structural traits of the netlist. For example, the presence of all gate types within a netlist might have a favorable implication on the security of a locking scheme.

Using both tests, we can evaluate the fundamental structural security properties of a scheme, thereby potentially revealing the cause for information leakage or a specific setting in which resiliency might be achieved.

Before attempting to introduce a learning-resilient scheme, it is first necessary to understand the reasons for the key-related leakage induced by existing locking schemes. Let us look at two aspects of logic locking: *the location selection* and *the introduced change*. The location selection defines how a scheme decides where the change is introduced. If the selection or the key-controlled change exhibits leakage, an entity (human or artificial system) is able to learn about the correctness of the key. Therefore, a locking scheme that *has no biased* location selection and introduces *unpredictable* changes (with respect to the key value) is learning-resilient. These two leakage points can sometimes be detected within a few iterations of the proposed test. To understand how we can utilize the tests for leakage detection, let us consider the examples in the following sections.

10.1.3   Test Application for XOR/XNOR-Based Locking

Let us analyze Random Logic Locking (RLL) with the introduced tests. In this scheme, the insertion location is randomly selected. However, RLL has been

---

[4] Evidently, any type of gate can be utilized within the test. Therefore, in practice, all primitive gate type must be evaluated to avoid a potential bias of a specific locking scheme.

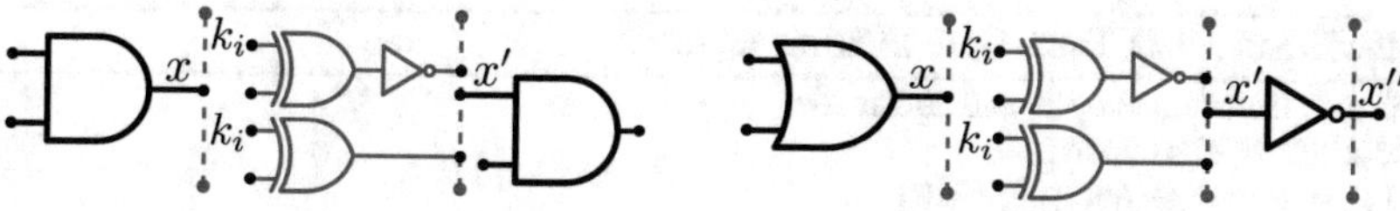

Fig. 10.2   ANT and RNT for XOR/XNOR-based locking

successfully attacked with ML-based attacks. Thus, some form of key-related information leakage must be provided through the introduced change.

10.1.3.1   ANT Observations

In the AND netlist test, **T** locks an AND-only netlist in each iteration using a randomly generated key, where XOR is inserted for 0 and XNOR for 1. After many iterations, **U** is able to learn that an XOR is correlated with the value 0 and XNOR with the value 1. This can be further understood through the example in Fig. 10.2a. Since **U** is aware that the original netlist exclusively contains AND gates in ANT, it is possible for U to isolate the exact location that is affected by the inserted key gate between two AND gates. In the provided example, it is evident that the value x entering a key gate must be buffered once processed to ensure the preservation of the original functionality. Thus, the attacker has to guess k_i such that it ensures this preservation, i.e., $k_i =? \Rightarrow x = x'$. Moreover, the operation mode of the scheme ensures that it is always true that $k_i = 0$ for XOR and $k_i = 1$ for XNOR.

Verdict Since **U** is able to learn about the key by isolating the limited spatial region around a key gate, RLL fails ANT and is considered vulnerable to learning-based attacks.

10.1.3.2   RNT Observations

In RNT, **T** locks netlists using a variety of gate types in each iteration. Therefore, there exists the possibility that an XOR key gate is placed in front of an existing inverter. This leads to observations where an XOR+INV is associated both with the key value 0 and the value 1. Hence, it is sometimes difficult for **U** to conclusively isolate the key gates, as visualized in Fig. 10.2b. Here, the correct cut-off line for the restoration of x is not easily identified. Consequently, **U** has to guess whether $x = x'$ or $x = x''$. In this setting, **U** is more likely to make *false guesses*.

Verdict In RNT, **U** is more likely to make a false guess compared to the ANT case. However, since the netlists in RNT consist of an evenly distributed number of gate types, these contradicting cases might not occur very often. Consequently, it is expected that the guessing accuracy in RNT is higher than 50%. Therefore, RLL fails RNT.

Algorithm 10.1: Twin-Gate locking scheme

Input: Activation key K and netlist Net
Output: Locked netlist
1   $T_{multi} \leftarrow \{AND, OR, \ldots, XOR\}$
2   $T_{single} \leftarrow \{NOT, BUF\}$
3
4   **for** $i = 0$ *to* $|K|$ **do**
5   $\quad n_{true} \leftarrow$ RandomSelection(Net)
6
      /* Create false node   */
7   $\quad$ **if** $TypeOf(n_{sel}) \in \{NOT, BUF\}$ **then**
8   $\quad\quad |\quad F_{types} \leftarrow \{t \in T_{single} \mid t \neq \text{TYPEOF}(n_{true})\}$
9   $\quad$ **else**
10 $\quad\quad |\quad F_{types} \leftarrow \{t \in T_{multi} \mid t \neq \text{TypeOf}(n_{true})\}$
11 $\quad$ **end**
12
13 $\quad f_{type} \leftarrow \text{RANDOMSELECTION}(F_{types})$
14 $\quad n_{false} \leftarrow \text{CREATENODEOFTYPE}(f_{type})$
15
      /* Assemble MUX node   */
16 $\quad m_{enc} \leftarrow \text{CREATEMUXFOR}(n_{true}, n_{false}, K[i])$
17 $\quad \text{REPLACENODES}(n_{true}, m_{enc})$
18 **end**
19 **return** Net

10.1.4   Test Application for Twin-Gate Locking

To present other aspects of the tests, we introduce the Twin-Gate scheme. As shown in Algorithm 10.1, Twin-Gate receives two inputs: the correct key and a target netlist. In the first step, the scheme prepares the set of multi-input (T_{multi}, line 1) and single-input gate types (T_{single}, line 2). For each key bit (line 4), Twin-Gate performs three steps: the random selection of a *true* node (line 5), the creation of a *false* node (lines 7–14), and the assembly of a replacement MUX (lines 16–17). The true node represents a node from the original netlist. The false node embodies the additional *twin* that is added as a pair to the true node. Depending on the type of the true node, Twin-Gate selects a suitable false node by ensuring that both nodes have the same number of inputs and a different type. For example, if the true node is of type BUF or INV, its twin can only be INV or BUF. In case the true node is a multi-input node, Twin-Gate randomly selects a type from the available multi-input gate types. Once the false node is created, a MUX is inserted and coupled to buffer the true and false node outputs. The current key input ($K[i]$) acts as a selection bit for the MUX to determine which gate output is forwarded. A locked example is shown in Fig. 10.3a. The true AND node is replaced by the pair (kg_0, kg_1). The correct gate output is selected by the key bit k_0.

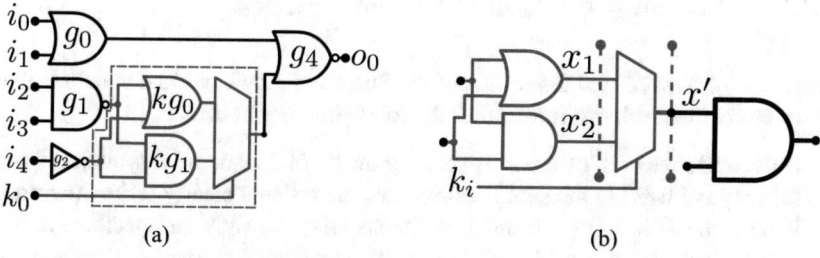

Fig. 10.3 Twin-Gate locking scheme. (**a**) Locked example. (**b**) ANT analysis

10.1.4.1 ANT Observations

In the case of ANT, Twin-Gate resolves the direct leakage problem of XOR/XNOR-based locking by avoiding a correlation between the inserted structure and the correct key bit. For example, let us consider the inserted key gates in Fig. 10.3b. Here, the scheme has replaced an original AND gate with the pair (OR, AND). Evidently, it is possible to isolate the exact gates that are part of the locking mechanism. In this case, identifying the correct key k_i is equivalent to guessing whether $x' = x_1$ or $x' = x_2$. By analyzing the samples individually, U is not able to learn whether AND or OR is the true node. However, it turns out that U is able to learn a correlation by analyzing the type and distribution of *all key gates* in the netlist. For example, after many iterations, U can observe that AND is never a false gate, as otherwise, two AND gates are fed to a MUX. Consequently, U can easily learn which key bit is correct for a given pair. The amplification of this effect is provided by the structural features of AND-based netlists that exclude AND gates as possible false gates.

Verdict Since the selection of false nodes depends on the structure of the original netlist, Twin-Gate fails ANT. However, as discussed in the next section, specific cases exist in which the scheme might exhibit learning resilience.

10.1.4.2 RNT Observations

The RNT setting creates well-balanced netlists in terms of the gate-type distribution. Hence, U has a disadvantage in guessing the correct key value, as every gate type is equally likely to be selected as a true or false node.

Verdict For the very specific case of random netlists, Twin-Gate passes RNT. However, in practice, circuits might not have a perfectly balanced amount of each gate type; thus, making it possible to identify potential false nodes.

10.1.5   Learning Resilience: Lessons Learned

Based on ANT, RNT, and the results of the SnapShot attack, we can conclude that a
learning-resilient scheme must fulfill the following requirements:

- **Independence of synthesis:** Its security must not depend on resynthesis.[5] If the
  security is linked to the synthesis process, the scheme clearly leaks information
  if resynthesis is not performed, as its security depends on specific synthesis
  transformations; which are often part of proprietary software. Moreover, this
  dependency has an important implication: the security of a locked design would
  not solely depend on the key.
- **Independence of design features:** The scheme-induced change and its location
  must not depend on the inherent structural of functional features of the target
  design.
- **Independence of key values:** The scheme-induced change and its location must
  not depend on the value of its driving key bit.

Hereby, the last requirement can be manifested in the following: the observations
that **U** makes during ANT and RNT must be either random (noisy data) or two
identical observations must suggest two different key values. The latter mirrors the
case when two identical images are labeled differently.

10.2   Structural Analysis Attack on MUX-Based Logic
Locking

The results of SnapShot as well as the example ANT/RNT analysis suggest that
achieving learning resilience in the form of XOR/XNOR-based locking is only
possible if the underlying netlist structure creates suitable locking conditions.
Thus, we focus the development of a deceptive scheme around MUX-based locking
approaches. MUX-based locking has a profound advantage over other key-gate-
based schemes; instead of inserting additional gates to manipulate the design, it
reconfigures the existing logic. Thereby, a MUX-based scheme always inserts the
same structures: multiplexers. However, through the analysis of common MUX-
based locking policies, we identified an important structural vulnerability that can
determine the correct key-bit value in an OL fashion.

To understand the working principle of MUX-based locking, let us consider the
example in Fig. 10.4a. The inserted MUX takes two inputs: the true (T) and false (F)
wire. T represents the original wire between g_1 and g_3, whereas F is a dummy wire.
The MUX is controlled via the key input k_0. Even though both values of the key bit
result in a functionally valid netlist, if F is not selected carefully, a simple analysis

---

[5] Resynthesis is often deployed to ensure the full integration of structural scheme-induced changes.

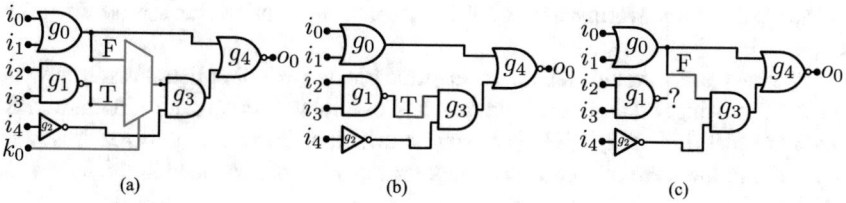

Fig. 10.4 Example: MUX-based locking and SAAM. (a) MUX Locking. (b) True wire selected. (c) False wire selected

---

Algorithm 10.2: SAAM: Structural Analysis Attack on MUX-based locking

---

Input: Locked netlist Net
Output: Extracted key E where $\forall e \in \{0, 1, X\}$
1 $E \leftarrow \{\emptyset\}$
2
3 **for** $i = 0$ *to* $|K|$ **do**
4      $\{n_1, n_2\} \leftarrow$ EXTRACTDIRECTINPUTSOFMUX(Net, $K[i]$)
5
6   **if** $(OUTSIZE(n_1) > 1) \wedge (OUTSIZE(n_2) > 1)$ **then**
7         $E[i] \leftarrow X$ // Set unknown value
8   **else if** $OUTSIZE(n_1) == 1$ **then**
9         $E[i] \leftarrow$ VALUEOFKEYFORMUXINPUT(n_1) // Out of n_1 is true wire
10   **else**
11        $E[i] \leftarrow$ VALUEOFKEYFORMUXINPUT(n_2) // Out of n_2 is true wire
12   **end**
13 **end**
14 **return** E

---

can easily identify the correct true wire as follows. Based on the locking example in Fig. 10.4a, the attacker can create two cases by removing the MUX and forwarding the respective wire for each value of the selecting key bit. If the MUX selects T as the correct wire (Fig. 10.4b), the resulting netlist displays no structural faults. However, if the MUX selects F as the correct wire (Fig. 10.4c), the wire T (output of gate g_1) remains an unconnected, dangling wire. Furthermore, this is *never* the case for the wire F, as it is randomly selected out of the set of original wires in the design. Hence, the gate that drives F (gate g_0) is always connected to some other gate as well (gate g_4). Therefore, we can formulate an attack based on the following: *the MUX input that remains unconnected when not selected is the true wire*. We summarize this vulnerability in the form of the OL Structural Analysis Attack on MUX-Based Locking (SAAM). An overview of the attack procedure is presented in Algorithm 10.2. SAAM takes the locked netlist as input and returns the extracted key, where each bit is set to 0, 1, or X (unknown). The attack works as follows. For each key bit, SAAM checks which input of the connected MUX has one output (lines 6 to 12). For that node, the corresponding key value is extracted.

SAAM fails at uncovering the key value if both input nodes have more than one output.

Note that even though SAAM is, in principle, a very simplistic attack, a wide range of existing MUX-based schemes has overlooked this major structural vulnerability [98, 124, 132, 152]. The structural faults uncovered by SAAM is further addresses in the form of a deceptive locking scheme as introduced in the following section.

10.3   Deceptive Multiplexer-Based Logic Locking

In the following, we transfer the concepts of MUX-based locking, SAAM, and learning resilience to the construction of a Deceptive Multiplexer Logic Locking (D-MUX). Hereby, the learning-resilience tests will be used to make a first evaluation of the resilience of the scheme in the context of Definition 10.1. Once constructed, we will re-evaluate the scheme with novel attack vectors that will shed light on vulnerabilities that must be further addressed to ensure a secure scheme.

The core functionality of D-MUX is built on specific *locking strategies* that ensure that each path through a MUX has the same probability of being true or false. In the following, we first introduce more details on the concept of locking strategies. Afterward, using these strategies, we compose D-MUX.

10.3.1   Locking Strategies

To achieve resilience against SAAM as well as dissolve the possibility of an educated guess determining the true wire, the wire selection during MUX insertion must be steered to avoid selecting single-output gates as candidates. Thus, we introduce multiple locking strategies that fulfill this criterion, following the visualization in Fig. 10.5. A single strategy S_i is defined by the following components:

- **Input node selection:** Selects two input nodes $\{f_i, f_j\}$. These nodes represent two gates that drive the inputs of one or multiple MUXs.
- **MUX configuration selection:** Selects two MUXs to be involved in a single locking iteration.
- **Key length selection:** Selects two one-bit key inputs $\{k_i, k_j\}$.
- **Output node selection:** Selects two output nodes $\{g^i, g^j\}$, where g^i is the output node of f_i if f_i drives one input of g^i in the target netlist.

    Moreover, the node f_i can be of the following types:

- **Single-output:** f_i drives only one output node g^i_1.
- **Multi-output:** f_i drives multiple output nodes $\{g^i_1, g^i_2, \ldots\}$.

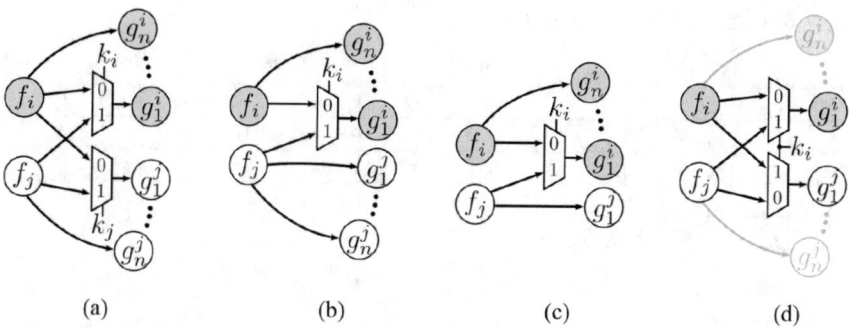

Fig. 10.5 D-MUX locking strategies. (**a**) S_1. (**b**) S_2. (**c**) S_3. (**d**) S_4

Using these components, we can derive several locking strategies: *Two-MultiOutTwoBitTwoMux* (S_1), *TwoMultiOutOneBitOneMux* (S_2), *OneMultiOutOneBitOneMux* (S_3), and *AnyOutOneBitTwoMux* (S_4). The strategies are discussed in detail in the following.

TwoMultiOutTwoBitTwoMux (S_1) S_1 selects two multi-output nodes $\{f_i, f_j\}$ (TwoMultiOut) to introduce a pairwise lock by using two individual key bits $\{k_i, k_j\}$ (TwoBit). Every key bit acts as selector signal for one particular MUX (TwoMux). A visualization is presented in Fig. 10.5a. One output node is selected for each input node, i.e., $\{g^i, g^j\}$, to select two initial *true paths*: $f_i \rightarrow g_1^i$ and $f_j \rightarrow g_1^j$. In the next step, two MUXs are placed between the selected input and output nodes to create four valid paths for all values of $\{k_i, k_j\}$. Since both input nodes originally have multiple outputs, one cannot determine which path is true or false, as all are equally valid. Furthermore, a path from an input to an output node does not have to exists at all. For example, if $k_i = 0$ and $k_j = 0$, input node f_j is neither driving g_1^i nor g_1^j. This is, however, valid, since f_j is a multi-output node and, therefore, remains connected even if not buffered through a MUX. All allowed configurations of S_1 are presented in Table 10.1a. The entries are interpreted as follows. For the input keys k_i and k_j, the nodes f_i and f_j are buffered to the nodes specified in the table. For example, if $k_i = 0$ and $k_j = 1$, the output of f_i is connected to g_1^i, while f_j is connected to g_1^j.

TwoMultiOutOneBitOneMux (S_2) S_2 selects two multi-output nodes $\{f_i, f_j\}$ (TwoMultiOut), thereby introducing a locking mechanism based on a single key bit k_i (OneBit) that drives a single MUX (OneMux). Afterward, the strategy randomly selects one output node of a randomly selected input node. For example, in Fig. 10.5b, S_2 selects f_i and one of its output nodes g_1^i. The MUX is placed between these two nodes, enabling the two configurations shown in Table 10.1b. Since both input nodes have multiple outputs in the original design, the case when one of the input nodes is not buffered to an output via the MUX remains valid.

Table 10.1 D-MUX strategy configurations

(a) S_1				(b) S_2			(c) S_3			(d) S_4		
k_i	k_j	f_i	f_j	k_i	f_i	f_j	k_i	f_i	f_j	k_i	f_i	f_j
0	0	g_1^i, g_1^j	$\emptyset$	0	g_1^i	$\emptyset$	0	g_1^i	$\emptyset$	0	g_1^i	g_1^j
0	1	g_1^i	g_1^j	1	$\emptyset$	g_1^i	1	$\emptyset$	g_1^i	1	g_1^j	g_1^i
1	0	g_1^j	g_1^i									
1	1	$\emptyset$	g_1^i, g_1^j									

OneMultiOutOneBitOneMux (S_3) This strategy selects only one multi-output node f_i (OneMultiOut) to be locked with a single key bit k_i (OneBit) that drives one MUX (OneMux), as presented in Fig. 10.5c. S_3 enables two configurations as shown in Table 10.1c. Note that S_3 differs from S_2 only in the fact that the output node must be selected from the multi-output input node. Thereby, regardless of which key bit is selected, both input nodes remain structurally connected.

AnyOutOneBitTwoMux (S_4) This strategy selects two input nodes $\{f_i, f_j\}$ from the set of all available nodes (AnyOut). Afterward, S_4 selects a single key bit k_i (OneBit) to drive two MUXs simultaneously (TwoMux). One output node is selected for each input node. The MUXs are configured to always forward the results of both input gates; otherwise SAAM is applicable in case a single-output gate is selected as input. The concept of S_4 is visualized in Fig. 10.5d. S_4 allows two configurations as presented in Table 10.1d. Regardless of the key-bit value, all node outputs form a valid path.

Note that driving each MUX in S_4 with an individual key is also a viable option. However, this specific strategy is susceptible to a simple reduction attack in case the input nodes are of the single-output type. For example, for $\{k_i, k_j\} = \{0, 1\}$, the node f_j remains unconnected, whereas for $\{k_i, k_j\} = \{1, 0\}$ the node f_i remains unconnected. Therefore, the attacker just has to guess whether $\{k_i, k_j\}$ is $\{0, 0\}$ or $\{1, 1\}$. This is equivalent to using S_4 with a single key.

10.3.2   Cost Model

The implementation cost for each locking strategy can be expressed in terms of the number of gates that are inserted per key K of length l, where $l = |K|$. The cost summary for all locking strategies is shown in Table 10.2 under the following assumptions:

- Every key input $k_i \in K$ is only used once.
- Every MUX is implemented with the same number of gates expressed as $|MUX|$. Typically, a multiplexer is implemented by using one inverter, two AND gates, and one OR gate, i.e., $|MUX| = 4$.

Table 10.2 Cost summary for D-MUX locking strategies

Locking strategy	Number of gates	Min. key bits per iteration
S_1	$l \cdot |MUX|$	2
S_2, S_3	$l \cdot |MUX|$	1
S_4	$l \cdot (2 \cdot |MUX|)$	1

Using this cost model, we can steer the final cost of D-MUX depending on *the available strategies*.

10.3.3 D-MUX Algorithm

By building on the presented locking strategies, we can assemble the overall D-MUX locking scheme as presented in Algorithm 10.3. The scheme takes the following inputs: a set of available strategies L_s, the correct key vector K, the original netlist Net, the maximum input node iteration variable I_{max}, and the maximum output node iteration variable O_{max}. I_{max} indicates the maximum number of reselections of input node pairs and O_{max} the maximum number of reselections of output node pairs for already selected input nodes. Since S_4 is always available to the locking procedure, it is not included in L_s. The final output is the locked netlist.

In the first step, D-MUX prepares two separate sets: all single-output (F_{single}) and multi-output (F_{multi}) nodes (lines 1 and 2). The key K is transformed to the list K_{list} (line 3) to facilitate the tracking of used keys. The main loop repeats until all keys have been used (line 5). Each iteration starts by randomly shuffling the available strategies (line 6) to ensure a random strategy selection in the next step. All strategies in L_s are iterated, and multiple checks are performed to ensure that enough key bits as well as input nodes of the desired type are available (lines 10 to 22). Once a candidate S_i is selected, the scheme proceeds with finding a valid pair of input and output nodes (line 27). If successful, the scheme performs the following steps based on the properties of S_i: retrieve the necessary key bits (set $K_{i,j}$, line 32) and couple all nodes through one or multiple MUXs (set $\{M_{i,j}\}$, line 33). Finally, all changes are stored in the netlist Net (line 34).

The input and output nodes selection must be carefully performed for all strategies to avoid the creation of combinational cycles. Therefore, D-MUX selects the nodes (line 27) according to the FindPairs function described in Algorithm 10.4. The function receives five inputs: the selected locking strategy S_i, the set of single-output nodes F_{single}, the set of multi-output nodes F_{multi}, and the maximum iteration variables I_{max} and O_{max}. The output consists of the found input ($\{f_i, f_j\}$) and output nodes ($\{g^i, g^j\}$) as well as an indicator (*done*) if the search has been successful. The function works as follows. First, two sets of nodes are assembled (F_1 and F_2) depending on the requirements of S_i (line 3 to 9). The sets are later utilized for the selection of input nodes. The main selection loop (line 14) is repeated until

---

Algorithm 10.3: D-MUX: **D**eceptive **MU**ltiple**X**er logic locking

---

Input: Available strategies L_s, key K, netlist Net, max input node iterations I_{max}, and max
　　　　output node iterations O_{max}

Output: Locked netlist

1 　$F_{single} \leftarrow$ EXTRACTSINGLEOUTPUTNODES(Net)
2 　$F_{multi} \leftarrow$ EXTRACTMULTIOUTPUTNODES(Net)
3 　$K_{list} \leftarrow$ TOLIST(K)　　　　　　　　　　// Convert key to list
4
5 **while** $|K_{list}| > 0$ **do**
　　　　// Select a candidate strategy
6 　　RANDOMSHUFFLE(L_s)
7 　　$fallback \leftarrow$ TRUE
8 　　$S_{sel} \leftarrow \emptyset$
9
10 　　**for** S_i *in* L_s **do**
　　　　　/* Enough keys available?　　　　　　　　　　*/
11 　　　**if** $S_i == S_1$ *and* $(|K_{list}| < 2)$ **then**
12 　　　　| **continue**
13 　　　**end**
14
　　　　　/* Enough nodes available?　　　　　　　　　　*/
15 　　　**if** $S_i \in \{S_1, S_2, S_3\}$ *and* $(|F_{multi}| < 2)$ **then**
16 　　　　| **continue**
17 　　　**end**
18
19 　　　$S_{sel} \leftarrow S_i$
20 　　　$fallback \leftarrow$ FALSE
21 　　　**break**
22 　　**end**
23 　　**if** $fallback$ **then**
24 　　　| $S_{sel} \leftarrow S_4$
25 　　**end**
26
　　　　/* Search for valid input/output nodes　　　　　*/
27 　　$\{\{f_i, f_j\}, \{g^i, g^j\}, done\} \leftarrow$ FINDPAIRS($S_{sel}, F_{single}, F_{multi}, I_{max}, O_{max}$)
28 　　**if** $!done$ **then**
29 　　　| **continue**
30 　　**end**
31
　　　　/* Apply selected strategy　　　　　　　　　　*/
32 　　$K_{i,j} \leftarrow$ GETANDREMOVEFROM(K_{list}, S_{sel})
33 　　$\{M_{i,j}\} \leftarrow$ COUPLETOMUXS($f_i, f_j, g^i, g^j, K_{i,j}$)
34 　　REGISTERTONETLIST($Net, \{M_{i,j}\}$)
35 **end**
36 **return** Net

---

two valid input and output nodes are found, or the termination condition (maximum
number of iterations) is reached. Each iteration selects two candidate input nodes
(f_i and f_j) from F_1 and F_2, respectively. In the next step, based on the input
nodes, the function tries to find two valid output nodes. The validity depends on

---

Algorithm 10.4: FINDPAIRS function

Input: Strategy S_i, set of single-output nodes F_{single}, set of multi-output nodes F_{multi}, max
     input node iterations I_{max}, and max output node iterations O_{max}
Output: Valid input nodes $\{f_j, f_i\}$, valid output nodes $\{g^j, g^i\}$, success indicator *done*

```
1  {F₁, F₂} ← {∅}
2
   /* Select correct nodes                                              */
3  if Sᵢ ∈ {S₁, S₂} then
4  |   {F₁, F₂} ← {F_multi, F_multi}
5  else if Sᵢ ∈ {S₃} then
6  |   {F₁, F₂} ← {F_multi, F_single}
7  else
8  |   {F₁, F₂} ← {F_single ∪ F_multi, F_single ∪ F_multi}
9  end
10
   /* Prepare input and output node references                          */
11 {fᵢ, fⱼ, gⁱ, gʲ} ← ∅
12 done ← FALSE
13
14 for iter_in = 0 to I_max do
       /* Select first and second input node                           */
15 |   {fᵢ, fⱼ} ← {RNDSEL(F₁), RNDSEL(F₂)})
16 |
       /* Select output nodes                                          */
17 |   for iter_out = 0 to O_max do
18 |   |   {Oⁱ, Oʲ} ← {OUTPUTSOF(fᵢ), OUTPUTSOF(fⱼ)}
19 |   |   {gⁱ, gʲ} ← {RNDSEL(Oᵢ), RNDSEL(Oʲ)}
20 |   |   {R₁, R₂} ← {ISINOUTCONE(fⱼ, gⁱ), ISINOUTCONE(fᵢ, gʲ)}
21 |   |
22 |   |   if (gⁱ != gʲ) and !R₁ and !R₂ then
23 |   |   |   done ← TRUE
24 |   |   |   break
25 |   |   end
26 |   end
27 |   if done then
28 |   |   break
29 |   end
30 end
31 return {{fᵢ, fⱼ}, {gⁱ, gʲ}, done}
```

---

three requirements (line 22): (1) the nodes are different, (2) f_j is not in the output
cone of g^i, and (3) f_i is not in the output cone of g^j. (2) and (3) prevent the creation
of combinational cycles. The node selection is controlled by a termination condition
that limits the search to a given number of iterations since a valid selection is not
always possible. The algorithm repeats the search if the selection was not successful.

Note that the selection always returns two output nodes, even though some
strategies require only one node. However, this does not impact the final result since
every node has at least one output. Finally, to avoid a simple identification of the

true and false nodes, the selected sets F_1 and F_2 never contain gates that assemble any of the inserted MUXs from previous iterations.

Scheme Variants As previously mentioned, D-MUX operates according to the set L_s. This set enables the creation of Generalized D-MUX (gD-MUX) and Enhanced D-MUX (eD-MUX) scheme variants. For eD-MUX, $L_s = \{S_1, S_2, S_3\}$. Hence, the scheme first resorts to deploying the less costly strategies. However, in case none of these strategies is viable because not enough gates or keys are available, the scheme *falls back* to using S_4 (Algorithm 10.3, lines 23 to 25). For gD-MUX, $L_s = \emptyset$. Consequently, the locking procedure always falls back to S_4.

Note that S_4 is always applicable, regardless of the available nodes in the netlist. Thus, the difference between gD-MUX and eD-MUX mirrors the cost discrepancy between two extremes: (1) the target netlist only supports S_4 (worst case) and (2) the target netlist supports all other strategies (best case). Due to the cost implications of each strategy, the two D-MUX variants significantly diverge in terms of overhead, as further discussed in Sect. 10.5.

10.4 Resilience Evaluation

This section evaluates D-MUX against all relevant attacks, including SAAM, SWEEP,[6] the learning-resilience test, and SnapShot. The evaluation was performed based on the ISCAS'85, ITC'99, and RISC-V Ariane core benchmarks listed in Table 9.2. All evaluations were performed on an AMD Ryzen 9 3900X processor with 64GB of RAM and an Nvidia GeForce RTX 2080 Ti graphic card. The following sections only discuss the evaluation results. The exact setup parameters for the SWEEP and SnapShot attack are discussed in Appendix D.5.

10.4.1 SAAM Evaluation

Due to the nature of the locking strategies within both D-MUX variants, all input gates of the inserted MUXs never result in dangling outputs. Hence, the scheme is fully resilient against SAAM.

---

[6] SWEEP is an oracle-less constant propagation attack that exploits the differentiating circuit characteristics that are expressed when a specific key value is hard-coded during resynthesis. This attack is particularly effective on MUX-based locking. More information is available in [4].

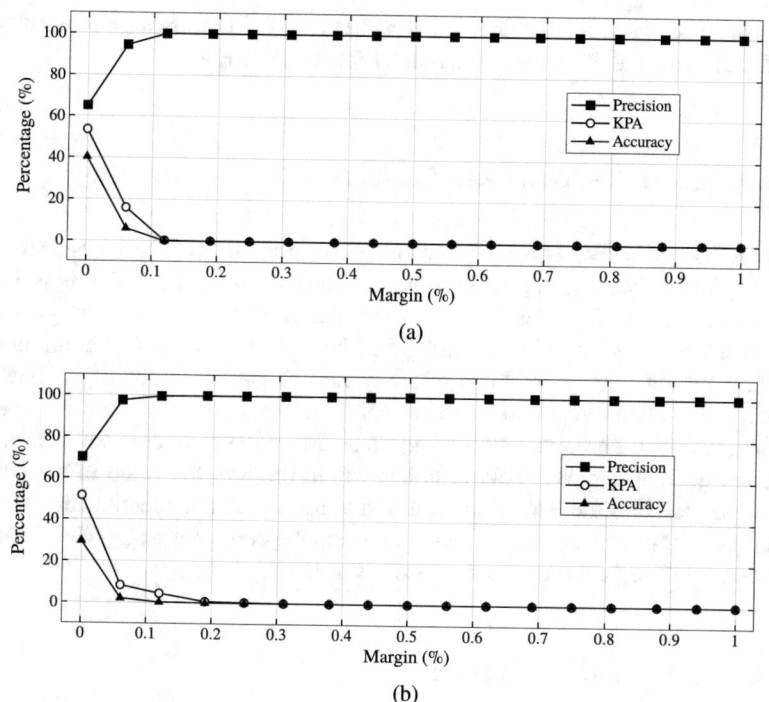

Fig. 10.6 SWEEP attack evaluation on D-MUX. (**a**) gD-MUX. (**b**) eD-MUX

10.4.2 SWEEP Evaluation

Metrics The SWEEP attack was evaluated through the accuracy, precision, and KPA metrics for a range of margin values m[7] as discussed in Appendix D.5.1. Hereby, the accuracy in SWEEP is determined using the whole key length, *regardless* of the number of X values, i.e., values that have not been conclusively guessed by the attack. In contrast, KPA only considers the bits that are *different* from X.

Results The average attack results across all benchmarks and margins are shown in Fig. 10.6. For $m = 0$, the average KPA is approx. 50% as expected. However, the accuracy is lower than 50% since its calculation takes X values into account. As soon as the margin value is increased ($m > 0$), both the KPA and accuracy drop to 0%, while the precision rises to 100%. This high precision occurs because SWEEP is not able to report any "guessed" key bits, i.e., all bits are marked as X. Hence, the high precision is, in this case, an indication that the attack has failed. Based on

---

[7] m adjusts the freedom of SWEEP to make "wild" guesses. By default, $m = 0$.

the evaluation, we can conclude that SWEEP is not able to extract a meaningful correlation from the data regardless of the acceptable margin.

10.4.3   Learning-Resilience Evaluation

In the AND netlist test, U is able to identify the inserted MUX gates by following each key input. Hereby, the only observation that U can make is that both inputs to the MUX are AND gates. However, the nature of the locking strategies in D-MUX disables the possibility to distinguish the true and the false gate driving the MUX. Even after many iterations of the game, the guessing accuracy is likely to remain 50%. Therefore, D-MUX passes ANT, thereby being evaluated as *possibly* learning-resilient. The same conclusion can be drawn in the random netlist test. The reason why D-MUX passes both tests lies within the way the value of the correct key is mapped into the netlist; instead of adding key-driven functionality, D-MUX reconfigures the existing one by combining existing combinational paths. Thereby, the same MUX structure is added regardless of the key value.

10.4.4   SnapShot Evaluation

We evaluated the D-MUX variants using both Generalized Set Scenario (GSS) and Self-Referencing Scenario (SRS), similar to the evaluation in Sect. 9.1. To adjust the SnapShot attack to the complexity of D-MUX, we adapted the Multi-Layer Percep-tron (MLP) and Convolutional Neural Network (CNN) models to accommodate for MUX-based locking. All setup details are available in Appendix D.5.2. The locality extraction for MUX-based locking is presented in Appendix D.5.3.

Results The average evaluation results for all attack scenarios, ML models, and D-MUX variants are presented in Fig. 10.7. For all attack configurations, the KPA is consistently around 50%. Hence, we can conclude that D-MUX is efficient in protecting against SnapShot, thereby forcing the attack to perform random guesses about the key. A detailed overview of the achieved KPA values for all benchmarks individually is available in Appendix D.5.2.

10.4.5   Security Requirements

D-MUX fulfills the learning-resilience criteria outlined in Sect. 10.1.5. The intro-duced locking strategies are completely independent of the synthesis process. This is empirically shown in the evaluations, *which did not include a resynthesis round after locking*. Furthermore, as discussed in Sect. 10.4.3, D-MUX passed the ANT and RNT tests, which indicates that the locking policy is not reliant on the structural

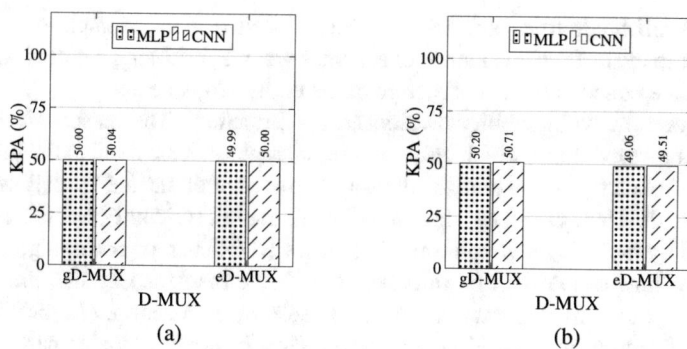

Fig. 10.7 SnapShot attack evaluation on D-MUX. (**a**) Average: GSS. (**b**) Average: SRS

netlist features. Finally, as D-MUX always inserts the same structures (MUXs), the key value is not correlated with the induced change.[8]

10.4.6 Security Challenges: Novel Attack Vectors

From the evaluation and analysis results, one might conclude that D-MUX is learning-resilient. And this conclusion is supported by all generated results. Both ANT and RNT were not able to conclusively state that a leakage exists. Note, however, that these tests cannot prove the absence of leakage. They can, however, prove its existence. The application of SnapShot and its extraction mechanism is designed to check for local leakage points, such as a relation between the key value and the local key-controlled gate structures. This relation is also checked through the theoretical tests, as they are performed by a human observer. Hereby, humans tend to focus on local observations as well. Thus, both the theoretical tests and the empirical evaluations have missed an important aspect of security in logic locking—a global bias. Even though a scheme might ensure that the inserted gate structures do not create a channel to the correct key value within *the locality of the key gates*, it does not mean it cannot be vulnerable to attacks that utilize a global knowledge about the hardware design. This aspect was recently introduced through the MuxLink attack [12] that is able to learn the structure and connectivity of the target design to predict correct links across the inserted key-driven multiplexers. Therefore, MuxLink does not guess the key value but rather predictions which link is more likely. Hence, MuxLink is a very effective attack that can comprise D-MUX as well as other MUX-based schemes. Note that the initial assumption of security in D-MUX—the absence of a predictable link between the inserted structures and

---

[8] Note that the existence of a global leakage is still not excluded. However, this challenge remains to be evaluated in future work.

the key—still holds true. Nevertheless, MuxLink is a great example of why it is important to evaluate the resilience of schemes when considering a global structural bias, i.e., the connectivity and structure of the entire target design.

Why was this vulnerability overlooked by SnapShot? The reason hides in the extraction methodology, which only captures changes that are local to the key gates. Thus, the extraction does not transfer any knowledge about the overall structural traits from the Hardware (HW) design to the ML model. Evidently, SnapShot can be improved by including a Graph Neural Network (GNN)-based processing.

Finally, how can D-MUX be improved to achieve resilience against link prediction? To resolve this vulnerability, the node selection procedure (Algorithm 10.4) must be steered to result in deceptive links. This, however, remains part of future research.

10.5   Cost Evaluation

Even though the cost of D-MUX is linear to the key length, the following evaluation gathers a realistic cost analysis using a concrete technology node. The evaluation is performed using the ISCAS'85 and ITC'99 benchmarks from Table 9.2. The logic synthesis is done using the Synopsys Design Compiler and the standard-performance cell library for the UMC 90 nm CMOS process operating at typical conditions (1 V, 25°C).

All benchmarks were locked with 64-bit keys for gD-MUX and eD-MUX, separately. To accommodate for locking and synthesis variations, the average cost is calculated across 20 synthesis rounds for 20 locked netlist instances. To extract a realistic average cost, we present the evaluations at three comparison points as follows. An overview of all cost-evaluation data points is available in Appendix D.6.

10.5.1   Optimum AT

The average Area-Power-Delay (APD) overhead for optimum Area-Timing (AT) points is shown in Fig. 10.8a. This evaluation quantifies the average difference between the *most efficient* synthesized implementations. Hereby, gD-MUX and eD-MUX exhibit an average area overhead of 73.38% and 47.26%, respectively. eD-MUX exhibits only 17.27% delay overhead compared to the 48.83% of gD-MUX. This is due to the fact that the AT optima of eD-MUX are typically closer in terms of T_{clk} to the original AT optima.

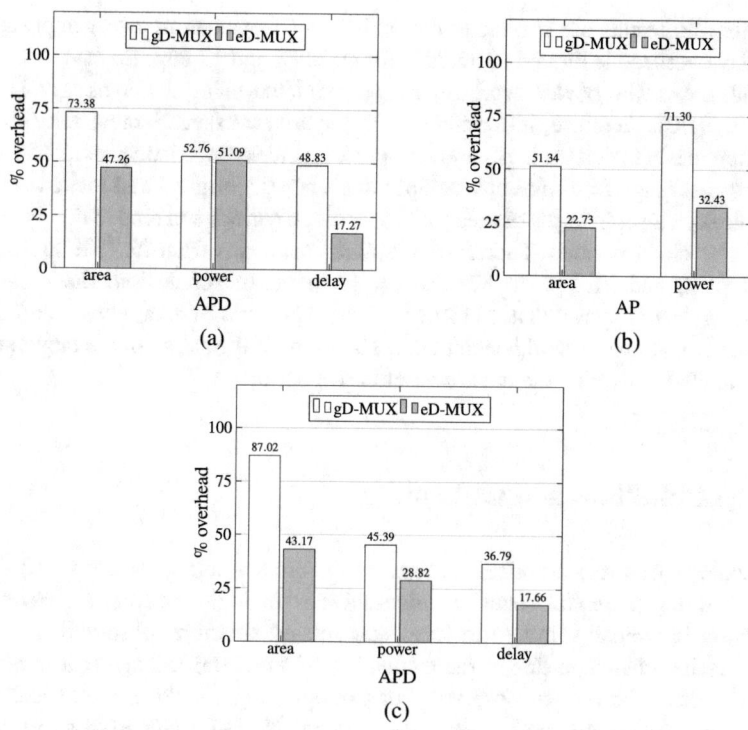

Fig. 10.8 D-MUX average cost evaluation. (**a**) Optimum AT. (**b**) Low performance. (**c**) High performance

10.5.2 Low Performance

The low-performance evaluation results are shown in Fig. 10.8b. The results summarize the cost difference for long clock periods[9] in which *no excessive optimizations* have been applied yet in the synthesis process, yielding a fair comparison. As before, gD-MUX exhibits a higher area overhead (51.34%) compared to eD-MUX (22.73%). The delay overhead is in both variants 0% since equivalent T_{clk} values are selected for all benchmarks.

10.5.3 High Performance

The average overhead at high-performance points is shown in Fig. 10.8c, summarizing the overhead for the lowest achieved T_{clk} (highest-performance designs). In

---

[9] Clock periods were selected to express comparison points at the flattened end of the AT curve.

this case, the results are similar to the optimum-AT evaluation. Most importantly, the delay overhead is limited to 36.79% for gD-MUX and 17.66% for eD-MUX.

In all cases, the power overhead is approximated using the Synopsys Design Compiler (DC); therefore, it correlates with the area overhead increase. Moreover, the difference between the high- and low-performance comparison provides insights into the cost range for different clock periods since the original and locked designs are typically closer in terms of area for longer clock periods and tend to diverge more for shorter clock periods. Therefore, the area cost spans from 22.73% to 43.17% for eD-MUX, and 51.34% to 87.02% for gD-MUX. In conclusion, the presented evaluation clearly shows that gD-MUX is costlier in terms of area, power, and delay compared to eD-MUX for all benchmarks. Therefore, if the nature of the target netlist allows it, eD-MUX offers the more cost-effective option.

10.6 Limitations and Outlook

The concept of the theoretical leakage tests offers a simple yet powerful way of uncovering elemental security vulnerabilities in logic locking. However, the evaluation is typically limited to local, key-related structural observations. Thus, an interesting challenge lies in the evaluation of structural leakage that is beyond the capability of a human observer. For example, logic locking might leak key-related information only on a large-scale, topological level. Evidently, analyzing the topology of netlists is not an easy task for humans. Furthermore, we have seen that the security of logic locking schemes is often predicated on the structural traits of the target netlists, i.e., the circuit family. Nevertheless, a comprehensive analysis of the existence of a global structural bias across a wide range of hardware designs still remains to be performed. A first indication that such a bias exists is given by the success of SnapShot as well as other ML-based attacks [170]. The results of such an analysis could provide important directives on what logic locking design should focus on if a general solution is infeasible.

10.7 Synopsis

The evaluation of SnapShot has uncovered an important structural vulnerability of XOR/XNOR-based locking schemes, thereby initiating the concept of learning resilience within the design of logic locking. By building on the lessons learned from the introduced attack, this chapter presented the concept of a learning-resilience test to provide a simple means of uncovering fundamental security flaws in logic locking. Using the theoretical concept, the design of a deceptive logic locking scheme with respect to ML-based attacks was introduced alongside an oracle-less attack on MUX-based locking. The security of the introduced scheme was further analyzed in the context of novel ML-driven attacks. With the presented work in this chapter, we established the cornerstones for the design of next-generation, learning-resilient logic locking in the era of machine learning.

Part V
New Directions

Chapter 11
Research Directions

Logic locking has been drastically investigated since its inception more than a decade ago. Its evolutionary landscape (see Chap. 5) clearly indicates that it has become increasingly difficult to find the right direction to further explore the capabilities of this protection approach [84]. To provide some starting points, in the following, we discuss potential further steps that are critical to a holistic security approach.

11.1 Improving Logic Locking

Beyond the improvement points discussed throughout the book, promising research directions are as follows.

11.1.1 Secure Key Storage

A fundamental assumption of the oracle-guided attack model is that the key cannot be physically inspected on the activated (in-silicon) chip. Unfortunately, as discussed in Chap. 4, from the technological point of view, that is not true. Hence, the key can be (thus far) accessed regardless of the capabilities of the underlying locking scheme. Therefore, enabling a secure key storage is a fundamental prerequisite to a secure, long-term locking approach within the oracle-guided model.

© The Author(s), under exclusive license to Springer Nature Switzerland AG 2023
D. Sisejkovic, R. Leupers, *Logic Locking*,
https://doi.org/10.1007/978-3-031-19123-7_11

11.1.2   Verifiable Security

Logic locking is an active protection mechanism. However, its impact is, thus far, unfortunately not verifiable nor measurable. Even though many measures of security w.r.t to logic locking have been proposed, these metrics have often focused on the "breakability" of schemes, not on the impact on, for example, the required effort to insert an intelligible Hardware (HW) Trojan. Such a measure could offer an invaluable tool for the design of integrity-protection methodologies. A much greater (a potentially more fruitful) research direction is having a formally verifiable logic locking approach. Hereby, the verifiability is directed toward its security implications, not the functional verifiability—as this is always possible (see Sect. 7.6.4).

11.1.3   Family of Circuits

HW is nowadays rapidly evolving in the form of multiple generations. This makes the problem of protecting its components even more complex, as generational similarities across multiple HW products could be exploited for reverse engineering and, ultimately, Trojan insertion. The success of physical attacks that can extract the locking key from a fabricated chip further exacerbates this issue. For a wider adoption, logic locking will have to handle such cases in the long term.

11.2   Untrusted IP and EDA Tools

As third-party Intellectual Property (IP) blocks are commonly used within the hardware design flow, protecting against infected IPs is an important and still unresolved problem. First promising approaches typically involve the utilization of machine learning to detect suspicious third-party IPs [73]. Moreover, different HW design stages are predicated on utilizing multiple electronic design automation tools. As these are mostly proprietary, closed-source software tools with full access to the designed HW, the tool-induced insertion of HW Trojans remains a real threat. One might argue that this problem can be resolved by simply performing an equivalence check between the outputs of multiple tools from different vendors. However, this scenario would assume that the inserted Trojans are not hidden in "don't care" states [74] and that the malicious adaptation of the tools is not extended to all vendors.

11.3   Security in Early Design Stages

As logic locking is typically deployed at the gate level, its protection mechanism does not shield against Trojan injections at higher design abstractions. Specifically, modern processor designs are often described using high-level design languages. Therefore, transferring the protection concepts of logic locking to these levels presents an interesting research opportunity. Moreover, also unintentional vulnerabilities in hardware can be exploited to compromise sensitive data. Recently, information flow analysis has been explored as a tool to detect and quantify leakage paths within hardware [62, 76, 117], thus enabling the identification of both intentional hardware Trojans as well as unintentional design mistakes at an early development stage (typically at register-transfer level) [135].

11.4   Security in Emerging Technologies

Emerging technologies and paradigms promise many advantages over the traditional HW characteristics. Therefore, their security aspects have to be carefully evaluated. A promising technology hides within neuromorphic computing. As this computing paradigm is still in the making, the community has the unique chance to introduce security into its developmental and production phase—an opportunity that was not available for CMOS. First investigations have already shown that many security vulnerabilities still persist in neuromorphic hardware [128, 172–174] and that the community must proceed with caution. Moreover, novel technologies can be utilized to drive security [26, 27]. For example, ion-sensitive field effect transistors can be deployed to generate secure keys from (bio)chemical information [115, 116].

11.5   Machine Learning for Security

As discussed in Chaps. 9 and 10, Machine Learning (ML) has become an important tool to evaluate the security of logic locking. However, the impact of ML goes far beyond integrity protection, thus offering fruitful challenges in Trojan detection, online monitoring, optimizations in synthesis and placement, vulnerability detection, and others [8, 58, 134, 183, 196, 213].

Chapter 12
Conclusion

Hardware (HW) is undoubtedly the most critical layer to security in electronic systems. However, the untrustworthy nature of the modern Integrated Circuit (IC) supply chain poses a great risk to the integrity of microelectronics. A tiny, intelligibly assembled, and correctly placed design modification can have catastrophic ramifications; from leaking sensitive user data to denial of service attacks on defense infrastructures—the range of possible attack vectors is nearly limitless. Therefore, protecting the integrity of HW throughout the IC supply chain is of paramount importance. This task, however, turns out to be exceptionally challenging, as the secret—the hardware itself—must remain functionally unchanged and structurally intact after fabrication, regardless of the deployed hardware-protection methodology. More than a decade of research was invested by the security community in designing a variety of protection schemes. Nevertheless, the increasing overspecialization of protection policies and inaccurate attack models has led to most solutions being limited to theoretical constructs without offering a tangible route for practical, secure HW development.

To address this challenge, this book aimed at narrowing the gap between theoretical concepts and industry-ready solutions for designing trustworthy hardware. Within this objective, we first introduced a generalized metric for measuring security with respect to logic locking. The metric captures both functional and structural implications of locking policies, thereby allowing for further customizations to support novel security properties. Furthermore, a security–cost trade-off case study was performed to evaluate the implications of the cost budget on the resilience of logic locking. Next, we introduced the design and implementation of an end-to-end logic-locking framework for the deployment of locking policies on complex, multi-module hardware designs. The framework was further extended with two protections schemes: Inter-Lock and Control-Lock. The former scales the security implications of any logic-locking policy across multiple hardware modules to counteract reverse engineering attacks. The latter protects critical control signals against the exploitation by software-controlled hardware Trojans. Both defense

© The Author(s), under exclusive license to Springer Nature Switzerland AG 2023
D. Sisejkovic, R. Leupers, *Logic Locking*,
https://doi.org/10.1007/978-3-031-19123-7_12

mechanisms were deployed and evaluated on silicon-proven RISC-V processor designs. Furthermore, the presented developments were commercialized, resulting in the first logic-locked processor, the "Made in Germany RISC-V" (MiG-V) core—a significant milestone in logic locking. Finally, we addressed the challenges of logic-locking design in the era of machine learning. Hereby, we introduced the SnapShot attack, which demonstrated the utilization of artificial neural networks to directly predict correct key bits from a locked netlist without the need for an activated IC. Coupled with a neuroevolutionary approach to automatically design suitable neural architectures, SnapShot demonstrated an average prediction accuracy of 82.6%, outperforming the state of the art for all evaluated benchmarks. By further analyzing the challenges of Machine Learning (ML)-resilient locking, we introduced the first test for identifying elemental security vulnerabilities that are exploitable by ML-driven attacks. Finally, by building on the results of the analysis, we developed a Multiplexer (MUX)-based ML-resilient locking policy and introduced an oracle-less attack that has uncovered a major fallacy in existing MUX-based schemes. To summarize, this book contributed toward transferring the potential of logic locking from theoretical constructs to practical applications, offering the concepts, tools, and metrics to build trustworthy hardware.

Appendix A
Notation Details

A.1 Graphic Representation of Logic Gates

The IEEE standard graphic symbols for logic functions are used to depict gate-level netlists in this book [77]. The legend of graphic symbols is presented in Fig. A.1.

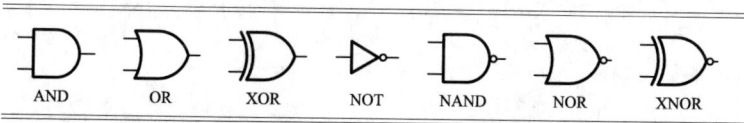

AND   OR   XOR   NOT   NAND   NOR   XNOR

Fig. A.1 Graphic symbols of logic gates

© The Author(s), under exclusive license to Springer Nature Switzerland AG 2023
D. Sisejkovic, R. Leupers, *Logic Locking*,
https://doi.org/10.1007/978-3-031-19123-7

Appendix B
Framework Details

This appendix includes further details about the logic locking framework described in Chaps. 7 and 8.

B.1 Parameters

The tool setup file consists of general and specific properties. The general properties define the overall framework setup, which is agnostic to the specifics of schemes. The specific properties adjust the scheme-specific configuration inputs. Parameters are defined in the following format: `the.parameter.key=the.parameter.value`. A description of the supported general configuration entries of the tool setup is listed in Table B.1. Specific properties of the Inter-Lock scheme (see Chap. 8) are listed in Table B.2. Note that a specific configuration is available for all implemented locking schemes.

B.2 Software Design Concept

As discussed in Chap. 7, the presented logic locking framework has been developed based on Object-Oriented Programming (OOP) design concepts. Thus, all framework components are modeled as individual classes, enabling a modular and reusable codebase. This allows for a simple integration process of new logic locking algorithms.

To visualize the implementation concept of the framework, a simple class diagram for the design of the Random Logic Locking (RLL) scheme is presented in Fig. B.1. Note that the class, parameter, and variable names have been simplified for this example. The model is based on two components: abstract (marked

© The Author(s), under exclusive license to Springer Nature Switzerland AG 2023
D. Sisejkovic, R. Leupers, *Logic Locking*,
https://doi.org/10.1007/978-3-031-19123-7

Table B.1 General parameters

Key	Option	Default value	Description
verbosity	true, false	true	If true, the framework outputs detailed information during execution
alg.type	rll, inter-lock, sll, d-mux, cas-lock	rll	Defines which scheme should be used for the locking procedure
module.names	*	*	List of module names that are included in the locking procedure
module.paths	*	*	List of file paths to all modules that are included in the locking procedure. The path order must match the order defined by module.names
output.dir	*	"./out"	Path to the output directory
blacklisted.inputs	*	"clk_i, rst_ni"	List of blacklisted inputs. These must not be affected by logic locking
external.dir	*	"./external"	Path to the external directory, which stores all external dependencies
pyverilog.cmd	*	"python3 [external.dir]/pyverilog/parser.py"	Defines the command to invoke the PyVerilog-based code parser
net.gen.cmd	*	"python3 [external.dir]/netlist_utils/enc_gen.py"	Defines the command to invoke the Python-based netlist generation script (used for Inter-Lock and Control-Lock)
randomize.gate.names	true, false	true	If true, all gate names are randomly generated
randomize.gate.names.len	*	4	Defines the length of the randomly generated gate names

(continued)

Table B.1 (continued)

Key	Option	Default value	Description
randomize.wire.names	true, false	true	If true, all wire names are randomly generated
randomize.gate.wire.len	*	3	Defines the length of the randomly generated wire names
rnd.node.prefix	*	l_n_	Naming prefix for randomly generated gates
rnd.wire.prefix	*	l_w_	Naming prefix for randomly generated wires
scheme.application.setup	*	"./scheme.application"	Path to the scheme application setup

Table B.2 Parameters for Inter-Lock

Key	Option	Default value	Description
interlock.output.prefix	*	int_o	Name of internal key wires
interlock.alg.type	*	rll	Defines which scheme is used for the per-module locking procedure within Inter-Lock
register.type	*	rising.edge.and.async.reset.low	Defines the type of registers implied by the constellation selection within the Inter-Lock hub
register.clock.name	*	clk_i	Defines the clock input name for the Inter-Lock registers
register.reset.name	*	rst_ni	Defines the reset input name for the Inter-Lock registers
register.suffix	*	_q	Defines the naming suffix for register insertion within the Inter-Lock hub
templates.dir	*	"../templates"	Path to the register templates. These are used to assemble the registers based on the register.type parameter
external.key	*	e_k	Array name of the external top-level key

(continued)

Table B.2 (continued)

Key	Option	Default value	Description
external.key.suffix	*	_e_k	Name suffix for defining external key inputs within the Inter-Lock hub
external.output.suffix	*	_int_o	Name suffix for defining internal key inputs within the Inter-Lock hub
total.key.suffix	*	_k	Name suffix for defining the total (per-module) key outputs within the Inter-Lock hub
interlock.hub.file.name	*	hub.v	Name of the Inter-Lock hub file
interlock.hub.instantiation.file.name	*	hub_instantiation.txt	Name of the Inter-Lock instantiations file

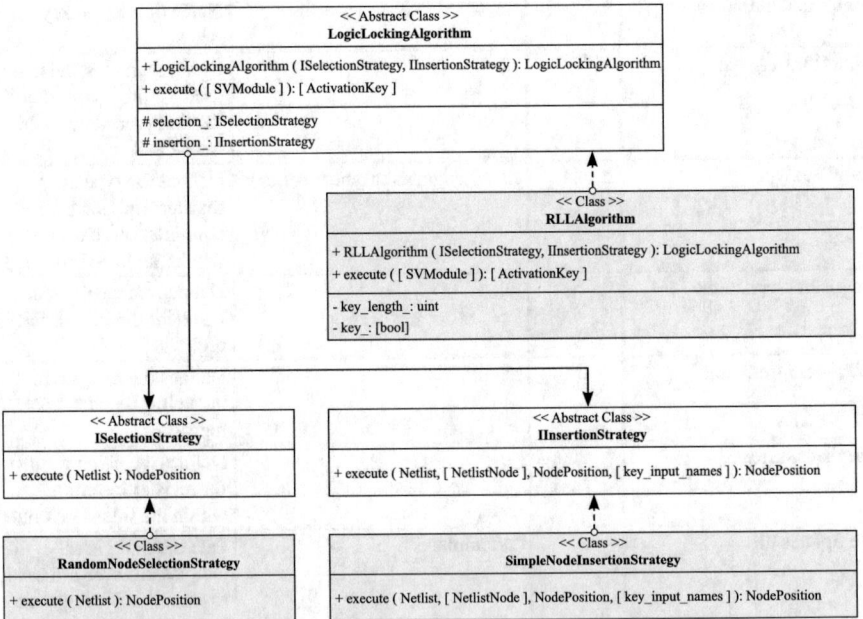

Fig. B.1 Example class diagram for RLL

white) and concrete classes (marked blue). The abstract classes model the overall implementation concept and the relations between the different components of the framework. The concrete classes model specific implementations of schemes and scheme components. Based on this concept, RLL is modeled as follows. The `LogicLockingAlgorithm` defines an abstract class of a logic locking scheme. The scheme operates on a given `SVModule` object. The `SVModule` object represents one instance of a preprocessed (System) Verilog gate-level netlist. In general, a logic locking scheme manipulates a netlist by inserting key gates or adjusting existing logic at specific locations. Thus, the `LogicLockingAlgorithm` references two additional abstract classes: the `ISelectionStrategy` and the `IInsertionStrategy`. The former models any class that defines a specific method for selecting suitable insertion locations. The latter represents classes that define particular key-gate insertion or manipulation methodologies. To implement RLL, the framework is extended by three concrete implementations:

- `RLLAlgorithm`: defines the concrete RLL algorithm. The class extends the abstract `LogicLockingAlgorithm` class, thereby having access to the selection and insertion objects.
- `RandomNodeSelectionStrategy`: defines a concrete selection strategy that ensures a random selection of key-gate insertion locations. This class extends the `ISelectionStrategy`.
- `SimpleNodeInsertionStrategy`: defines a concrete insertion strategy that is able to embed XOR/XNOR gates into selected locations in the netlist. This class extends the `IInsertionStrategy`.

The `RLLAlgorithm` is deployed via the `execute` function. This function receives a vector of `SVModule` references and returns a vector of `ActivationKey` objects. Other implementation-specific variables are defined in the concrete classes. By following this pattern, the framework can easily be extended with novel logic locking policies. Similar concepts have been developed for the design and implementation of all other framework components as well.

B.3 Inter-Lock Hub

This section showcases a concrete example of a generated hub. The initial Inter-Lock configuration is shown in Listing B.1 (equivalent to the setup in Sect. 8.1.1.1).

Listing B.1 Example of an Inter-Lock configuration

```
1  # --- Inter-Lock configuration ---
2  m1 8
3  m2 m3:2:f m4:3:t 12
4  m3 9
5  m4 4
```

Based on this scheme application setup and the tool configuration, after locking, the framework automatically generates the Inter-Lock hub as shown in Listing B.2. The hub is constructed as a separate Verilog module, receiving all external and internal keys as inputs. In this example, the internal keys are marked with the suffix _int_o. However, this can be adjusted in the tool setup file. Based on the constellation parameters, the hub constructs the final input key for each module (Lines 46–50). If necessary, internal keys are buffered through registers (Lines 37–43). For convenience, all external and internal keys as well as the constellation setup are stored in the hub's header (Lines 1–15). Note that this information must be removed before the design is handed to untrusted parties.

Listing B.2 Example of an Inter-Lock hub

```
1    // --- Correct (total) keys ---
2    // logic [7:0] m1_k = 8'b 00110101;
3    // logic [11:0] m2_k = 12'b 000100111110;
4    // logic [10:0] m3_k = 11'b 00101000010;
5    // logic [6:0] m4_k = 7'b 1101101;
6    // --- Correct external keys ---
7    // logic [7:0] m1_e_k = 8'b 00110101;
8    // logic [11:0] m2_e_k = 12'b 000100111110;
9    // logic [8:0] m3_e_k = 9'b 010100010;
10   // logic [3:0] m4_e_k = 4'b 1101;
11   // --- Inter-Lock configuration ---
12   // m1 8
13   // m2 m3:2:f m4:3:t 12
14   // m3 9
15   // m4 4
16   module interlock_hub (
17       // --- Register controls ---
18       input logic clk_i,
19       input logic rst_ni,
20       // --- External key inputs ---
21       input logic [7:0] m1_e_k,
22       input logic [11:0] m2_e_k,
23       input logic [8:0] m3_e_k,
24       input logic [3:0] m4_e_k,
25       // --- Interlock generated keys ---
26       input logic [4:0] m2_int_o,
27       // --- Final (total) keys ---
28       output logic [7:0] m1_k,
29       output logic [11:0] m2_k,
30       output logic [10:0] m3_k,
31       output logic [6:0] m4_k
32   );
33       // Define internal signals
34       logic [2:0] m2_int_o_m4_q;
35
36       // Register definition
37       always_ff @(posedge clk_i or negedge rst_ni) begin
38           if (~rst_ni) begin
```

```
39                    m2_int_o_m4_q <= 3'b0;
40            end else begin
41                    m2_int_o_m4_q <= m2_int_o[2:0];
42            end
43        end
44
45        // Define the wiring
46        assign m1_k = m1_e_k[7:0];
47        assign m2_k = m2_e_k[11:0];
48        assign m3_k =
                {m2_int_o[1],m3_e_k[8:2],m2_int_o[0],m3_e_k[1:0]};
49        assign m4_k = {m4_e_k[3],m2_int_o_m4_q[2:1],m4_e_k[2],
50            m2_int_o_m4_q[0],m4_e_k[1:0]};
51    endmodule
```

After all output files have been collected, the hub can be instantiated in the
top module of the locked design. All other wiring is routed through the design
hierarchy as discussed in Sect. 7.5. To simplify the process of key splitting and
hub integration, the framework additionally prepares a code template containing
the necessary instantiations, as shown in Listing B.3.

Listing B.3 Example of an Inter-Lock hub instantiation

```
1    // EXTERNAL KEY
2    logic [32:0] e_k = 33'b
         001101010001001111100101000101101;
3
4    logic [3:0] m4_e_k;
5    assign m4_e_k = e_k[3:0];
6
7    logic [8:0] m3_e_k;
8    assign m3_e_k = e_k[12:4];
9
10   logic [11:0] m2_e_k;
11   assign m2_e_k = e_k[24:13];
12
13   logic [7:0] m1_e_k;
14   assign m1_e_k = e_k[32:25];
15
16   // Interlock wires
17   logic [4:0] m2_int_o;
18
19   // Internal total key wires
20   logic [7:0] m1_k;
21   logic [11:0] m2_k;
22   logic [10:0] m3_k;
23   logic [6:0] m4_k;
24
25   // Hub instantiation (assumed clk/reset
         signals:clk_i/rst_ni)
26   interlock_hub i_interlock_hub (.*);
```

Removal Resilience Since the hub is a separate module, it can be vulnerable to removal attacks. Therefore, it is necessary to make sure a flattening process is deployed after integration to distribute the wiring in the design hierarchy. Nevertheless, this step is typically performed after any form of locking, as mentioned in Sect. 7.6.5.

Appendix C
Logic Locking and Machine Learning Details

This appendix is dedicated to introducing the basic concepts of machine learning with respect to the discussion and evaluations presented in Chap. 9.

C.1 Deep Learning and Neural Networks

A major role in the domain of deep learning (a subset of machine learning) is played by Convolutional Neural Networks (CNNs)—a specific type of Artificial neural networks (ANNs) that were initially designed for 2-dimensional convolutions inspired by the biological process of the visual cortex of animals [96]. Ever since their introduction, CNNs have pioneered the area of object detection and classification, specifically due to the vast amount of available data as well as fast, parallel computation.

The superiority of CNNs compared to traditional shallow ANNs comes from their ability to subsequently extract and utilize meaningful data features, specifically in unstructured image data. A general blueprint of a CNN is presented in Fig. C.1. The input to the network is an image. The output is typically a classification prediction. Structurally, a CNN consists of two major components: *feature extraction* and *classification*. The former is composed of the typical CNN layers that we refer to as *internal layers*.[1] The latter is based on a fully connected neural network, i.e., a network in which every hidden neuron is connected to all neurons in the preceding and succeeding layers.[2]

The internal layers are typically implemented as a sequence of *convolution* and *pooling* layers. Convolution layers extract features from the given data input by detecting local feature conjunctions that are typically further processed with an

---

[1] Internal layers are sometimes also referred to as hidden layers.

[2] Fully connected layers are often referred to as dense layers.

© The Author(s), under exclusive license to Springer Nature Switzerland AG 2023
D. Sisejkovic, R. Leupers, *Logic Locking*,
https://doi.org/10.1007/978-3-031-19123-7

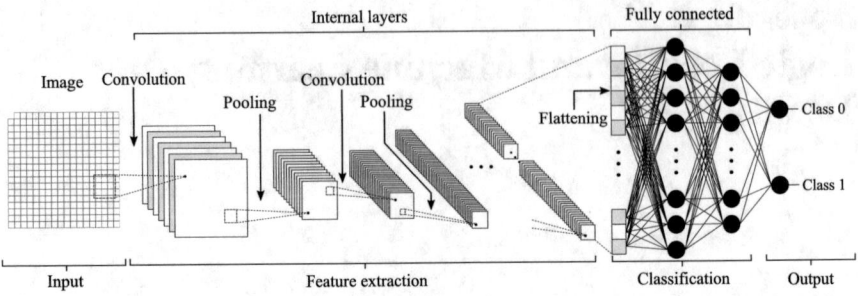

Fig. C.1 Architecture of a convolutional neural network

activation function to increase the non-linearity. Pooling layers progressively reduce the spatial size of the internal network representation to reduce the total amount of computation and data in the network.

After all features have been extracted through a sequence of convolution and pooling layers, the flattening layer transforms the data into a 1-dimensional array that is further forwarded as input to the classification component. With the composition of the feature extraction and classification, CNNs offer a powerful tool to *classify* images based on the automatically extracted *features*.

Another important neural network is known as Multi-Layer Perceptron (MLP). This network type implements a shallow, feedforward neural network that processes flattened input vectors rather than 2D or 3D data. Compared to CNNs, MLPs lack the ability to efficiently extract multiple feature dimensions. More details can be found in [17, 96, 103].

C.2 Genetic Algorithms

Genetic algorithms (GAs) belong to the class of evolutionary algorithms that embody meta-heuristic optimization techniques inspired by biological evolution [57]. Typically, GAs operate on a population-based model in which the individual solutions compete based on their *fitness* (quality) that is evaluated using a fitness function. Throughout multiple generations, solutions in the population are iteratively refined to exhibit higher fitness values through a subsequent application of the genetic operators, including selection, crossover, and mutation. A key component of GAs is the representation of a solution within the evolutionary algorithm. This manipulable encoding is referred to as *genotype*. The actual manifestation of the genotype is known as *phenotype*. A key feature of evolutionary algorithms is their ability to generate good solutions through a black-box approach, i.e., without the need for any domain-specific knowledge to perform the optimization. This feature is embodied by the decoding mechanism that translates the genotype into the phenotype representation of a solution.

C.2.1   Neuroevolution

Neuroevolution captures the evolution of neural networks with the help of evolutionary algorithms [7]. Neuroevolution can be utilized to evolve the network structure and weights. Note that neuroevolution is only one of many available Neural Architectural Search (NAS) methods [59] that belong to AutoML—a research area that focuses on methods to make machine learning more efficient, automated, and easy to use. More details are available in [67].

C.3   CNN Architecture Evolution

The CNN architecture evolution is implemented based on a neuroevolutionary process as presented in Fig. C.2. This process utilizes a genetic algorithm to search for suitable CNN architectures for the given prediction problem (Fig. C.2a). Through multiple generations, the GA adapts its solution pool through the selected genetic operators, advancing the overall quality of solutions. A visualization of the operation principle of the applied genetic operators is available in Fig. C.2c. As the GA operates in a black-box fashion, it requires a suitable *fitness evaluation function* to guide the heuristic search (Fig. C.2b). Thus, it is required to define how a solution is represented (genotype), decoded (phenotype), and how its quality is measured.

C.3.1   Genotype

The genotype embodies a domain-independent representation of a solution. To define a genotype for the CNN architecture evolution, we need to understand how a CNN is constructed. As detailed in Appendix C.1, a CNN architecture consists of two components: the feature extraction and the classification. These components are described by a set of properties, such as the number of layers, layer types, input/output dimensions, and others. Some of these properties must remain fixed to provide a basis for valid CNN architecture construction. Therefore, we define a fixed architectural frame that is constructed with a fixed input convolutional layer and the last two layers (a fully connected layer and the final output layer). This fixed frame is based on common architectural features, where the first internal layer is always a convolutional layer, and the classification is typically implemented in the form of two fully connected layers. Within this frame, the evolutionary process can fine-tune the architectural traits by altering the *number* and *type* of the internal CNN layers. To make these variables accessible to the GA, the genotype is encoded in the form of a *bitstring* of size $2 \cdot L$, where the first L bits denote the presence of the individual internal layers ($[x_1^p, \ldots, x_L^p]$) and the last L bits the layer types

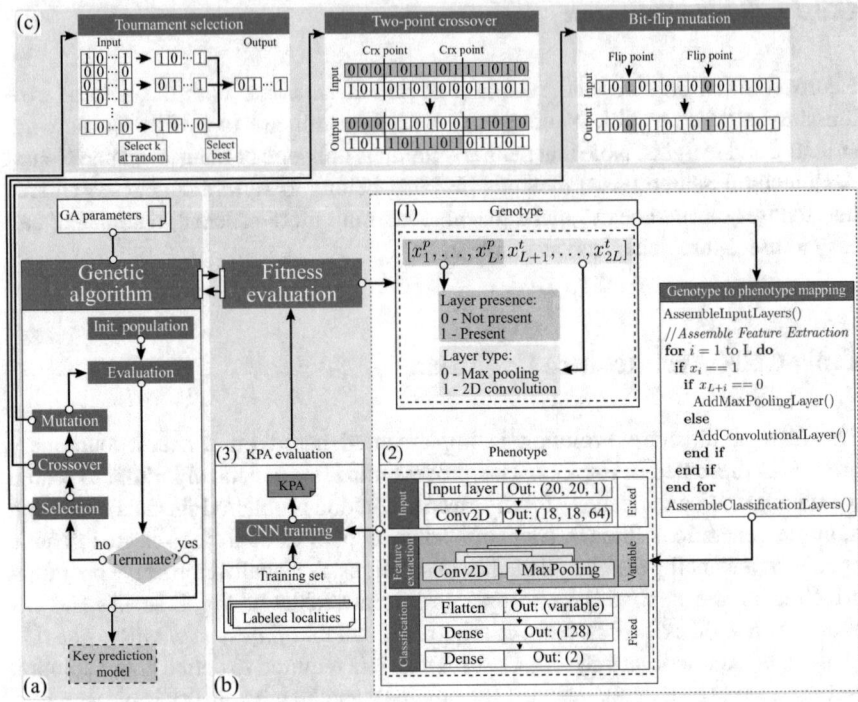

Fig. C.2 CNN architecture evolution: (**a**) genetic algorithm, (**b**) fitness function evaluation, and (**c**) examples of genetic operators

$([x^t_{L+1}, \ldots, x^t_{2L}])$. For example, if $x^p_i = 1$, the i-th internal layer is present and its type is defined by x^t_{i+L} (Fig. C.2b (1)).

C.3.2   Phenotype

The introduced genotype is a blueprint, i.e., a set of instructions that define how an architecture is constructed. To enable the fitness evaluation of a solution, its genotype must be decoded into a concrete CNN instance, i.e., the phenotype. This is performed by processing the genotype from left to right, thereby placing the selected internal layers within the fixed frame (Fig. C.2b (2)). The given input and output dimensions of the fixed layers are selected based on the input image size and the selected hyperparameters.

C.3.3 KPA Evaluation

The constructed CNN architectures must be evaluated according to a selected measure. In terms of the SnapShot attack, we evaluate the CNN architectures according to the achieved Key Prediction Accuracy (KPA) (see Sect. 9.3.1) on a given training set of labeled localities (Fig. C.2b (3)). Afterward, the achieved KPA is assigned as the fitness to the respective solution.

Appendix D
Evaluation Details

This appendix presents further details on evaluations that have been discussed in this thesis.

D.1 Impact of Cost Budget on Security

The results of the attack-resilience evaluation for the Boolean Satisfiability Problem (SAT) attack are shown in Table D.1. The entries in the percentage columns denote the used key length for a particular benchmark and area overhead. The resilience rate indicates the total percentage of *unbroken* locked circuits. The results show that the percentage of unbroken circuits steadily increases with an increasing locking budget.

A similar evaluation using the partial-break attack [175] and the Path-Sensitization Attack (PSA) [129] was performed. Hereby, the same setup is assumed as in Sect. 6.5.1. Both attacks belong to the oracle-guided model. Compared to the SAT attack, these attacks are capable of returning a partial key. The results are presented in Figs. D.1 and D.2. It can be observed that the percentage of retrieved key bits is decreasing for higher overheads. Thus, the evaluation conforms with the conclusions from Sect. 6.5.

D.2 Area-Timing Plot

The Area-Timing (AT) plot is a standard tool to evaluate the post-synthesis area-timing trade-off. One demonstrative example is presented in Fig. D.3. The plot visualizes how the logic-synthesis process impacts the design area for a range of clock periods (T_{clk}), yielding the typical AT curve. Hereby, the area is represented

© The Author(s), under exclusive license to Springer Nature Switzerland AG 2023
D. Sisejkovic, R. Leupers, *Logic Locking*,
https://doi.org/10.1007/978-3-031-19123-7

Table D.1 Resilience against the SAT attack per area overhead. The shown values represent the key length. The marked (gray) fields indicate unbroken locked circuits

IC	#gates	Area overhead					
		50%	60%	70%	80%	90%	100%
c432	161	80	96	112	128	144	160
c499	203	101	121	141	162	182	202
i4	338	169	203	237	270	304	338
c880	384	192	230	268	306	345	383
c1355	547	273	328	382	437	491	546
c1908	881	440	528	616	704	792	880
i9	1035	518	621	725	828	932	1035
ex5	1055	528	633	739	844	950	1055
c2670	1194	597	716	835	954	1074	1193
i7	1315	658	789	920	1052	1184	1315
c3540	1670	835	1001	1168	1335	1502	1669
k2	1815	908	1089	1271	1452	1634	1815
dalu	2298	1149	1379	1609	1838	2068	2298
c5315	2308	1154	1384	1615	1846	2076	2307
i8	2464	1232	1478	1725	1971	2218	2464
c7552	3513	1756	2107	2458	2810	3161	3512
seq	3519	1760	2111	2463	2815	3167	3519
ex1010	5066	2533	3040	3546	4053	4559	5066
apex4	5360	2680	3216	3752	4288	4824	5360
des	6473	3237	3884	4531	5178	5826	6473
	Resilience	15%	19%	35%	30%	45%	55%

in Gate Equivalent (GE) and the clock period in ns. The term GE stands for a unit of measure which allows specifying the area of digital circuits in a manufacturing-technology-independent form. One GE is the area of one 2-input drive-1 NAND gate. Therefore, the translation to GE is done by dividing the technology-specific area by the area of the mentioned NAND gate. The GE unit enables a fair comparison of designs regardless of the underlying technology node.

The AT plot offers a cost analysis of a design for low- and high-performance points. The low-performance analysis captures the design points for sufficiently high values of T_{clk}, where the logic-synthesis tool has not deployed any highly specific and, typically, area-costly optimizations. Thus, this AT-spectrum end offers a fair comparison between a set of designs. The high-performance points, on the other hand, represent the designs for the lowest-possible T_{clk} values, thus capturing the fastest implementations.

Another important point in the AT curve of any design is the AT optimum. This point represents the Pareto-optimal point with the most efficient design implementation, i.e., the point with the lowest $area \cdot T_{clk}$ value. Note that less-efficient design points, i.e., the ones not placed on the Pareto-optimal curve, are typically excluded from the visualization.

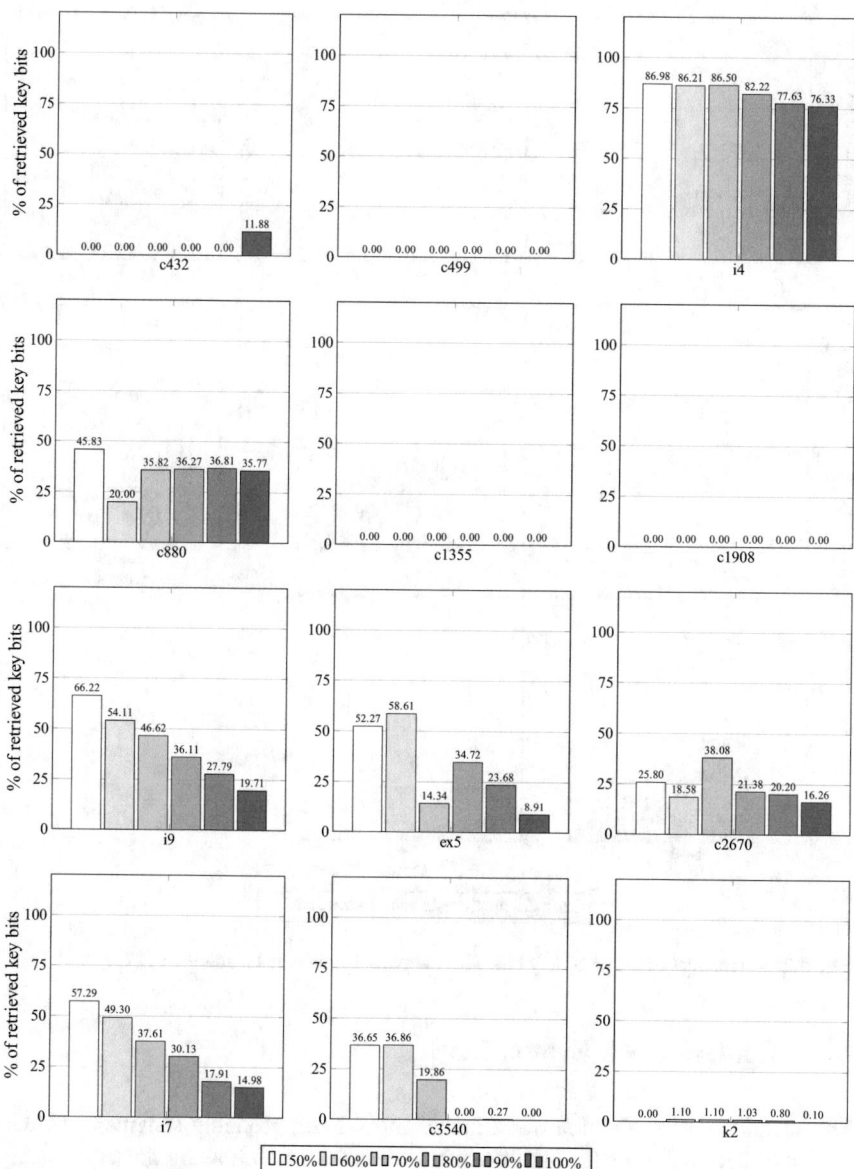

Fig. D.1 Percentage of retrieved key bits with the partial-break attack and PSA (1)

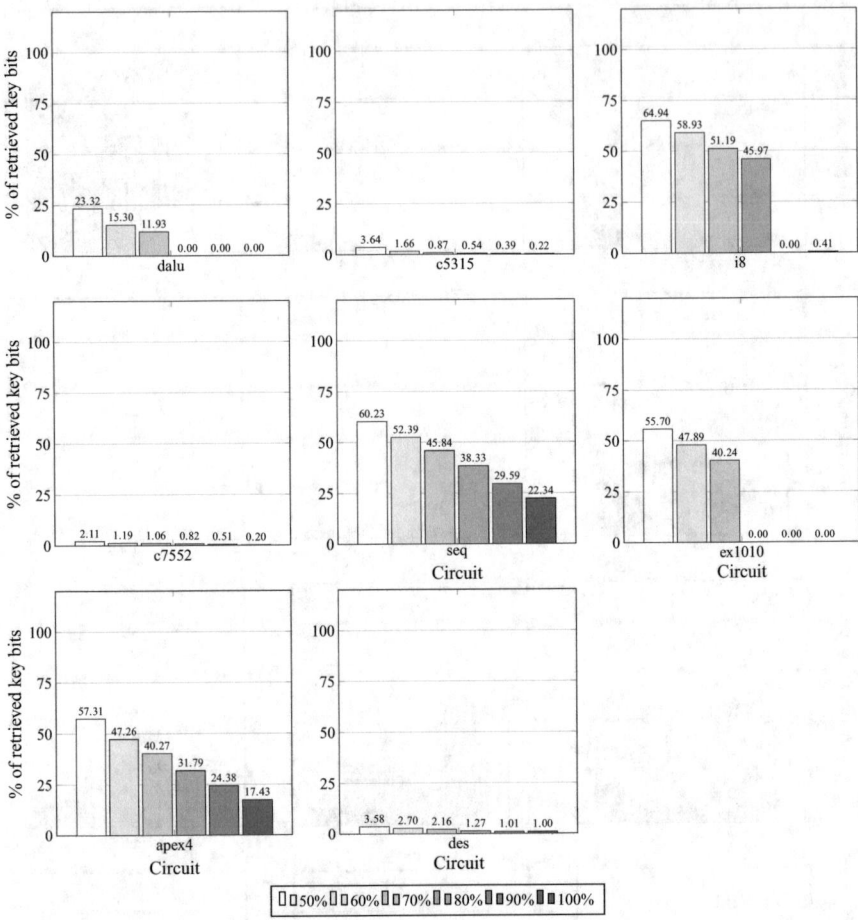

Fig. D.2 Percentage of retrieved key bits with the partial-break attack and PSA (2)

D.3   Control-Lock Evaluation

The security–cost evaluation of various Control-Lock grouping schemes was discussed in Sect. 8.2.5 for the high-performance case. The following presents further details on the AT-optimum, low-performance, and high-performance evaluation.

To extract all the necessary data, an AT plot for both modules was constructed, as shown in Figs. D.4 and D.5. The lowest achieved clock periods for the original design (without Control-Lock implications) are $T_{clk} = 0.20$ ns and $T_{clk} = 1.30$ ns for the decoder and Arithmetic Logic Unit (ALU), respectively. In both cases, the starting clock period was set to a sufficiently high value to result in a reasonable AT curve. All AT optima are highlighted in both plots.

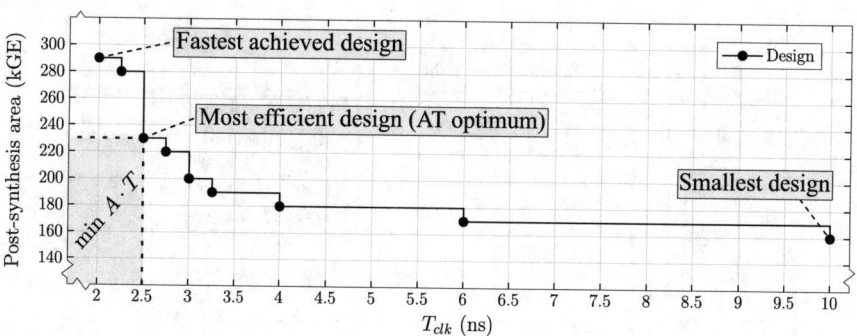

Fig. D.3 Area-timing plot

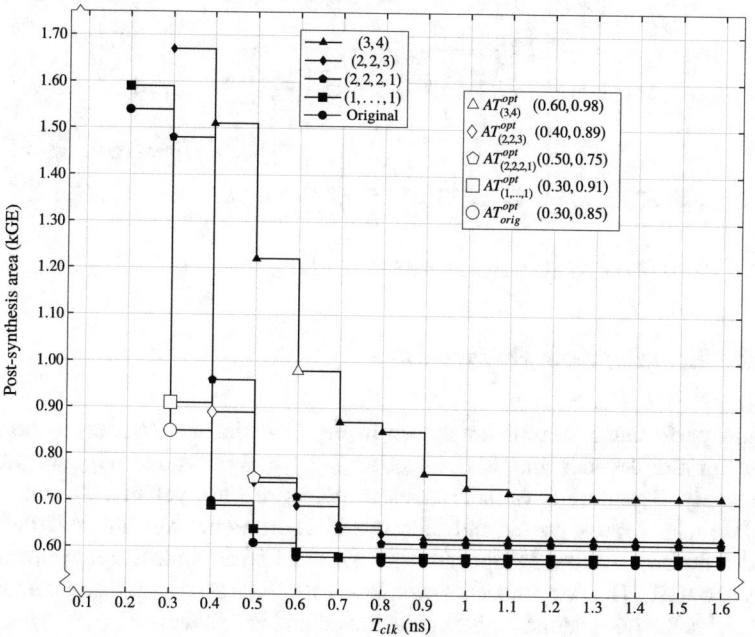

Fig. D.4 Control-Lock AT evaluation for the RI5CY decoder

D.3.1 Results: Optimum AT

The cost impact of all evaluated schemes for AT-optimum design points is shown in Fig. D.6. For the decoder, the area overhead ranges from -11.76% to 15.29%, the power overhead is limited to a maximum of 7.98%, and the delay ranges from 0.00% to 100.00%. The overhead range tends to be limited in magnitude for larger designs. Thus, for the ALU, the maximal area, power, and delay overhead is 12.59%, 0.02%, and 6.67%, respectively.

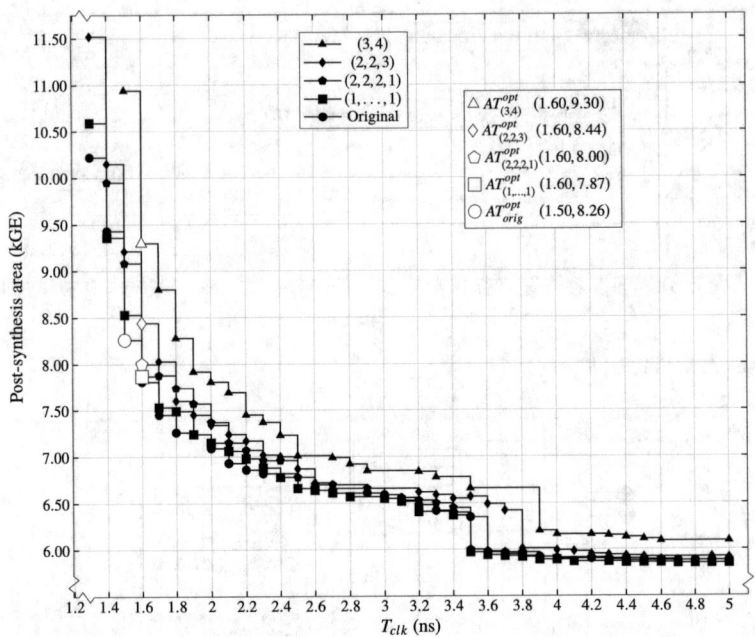

Fig. D.5 Control-Lock AT evaluation for the RI5CY ALU

D.3.2   Results: Low Performance

The low-performance results are shown in Fig. D.7. For the evaluation, the clock period for the decoder and ALU is set to $T_{clk} = 1.60$ ns and $T_{clk} = 5.00$ ns, respectively. The design variants at these points are not yet heavily optimized. Thus, the relative area, power, and delay overhead difference remain limited. For the decoder, the area overhead ranges from 1.75% to 24.56%, while the power overhead peaks at 8.91%. The cost impact is even lower for the ALU. Here, the maximal area overhead is 3.57%, and the highest power overhead is 2.09%.

D.3.3   Results: High Performance

The high-performance results have been discussed in Sect. 8.2.5. For convenience, the results are additionally presented in Fig. D.8.

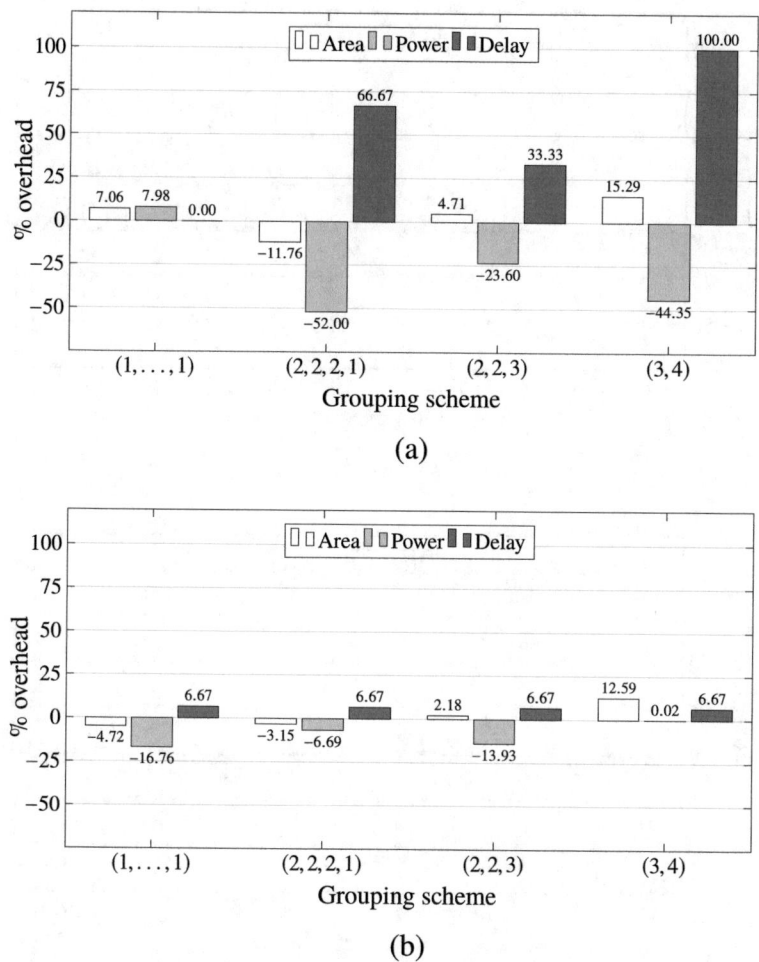

Fig. D.6 Control-Lock **AT-optimum** cost evaluations for the RI5CY (**a**) decoder and (**b**) ALU

D.4   ML Model Design

The following details the MLP and CNN parameters involved in the evaluation of the Machine Learning (ML)-driven SnapShot attack, as discussed in Sect. 9.3.

D.4.1   MLP Design Parameters

The evaluated MLP architecture is constructed as follows. All layers are fully connected. The input layer consists of 400 units to fully capture the length of the

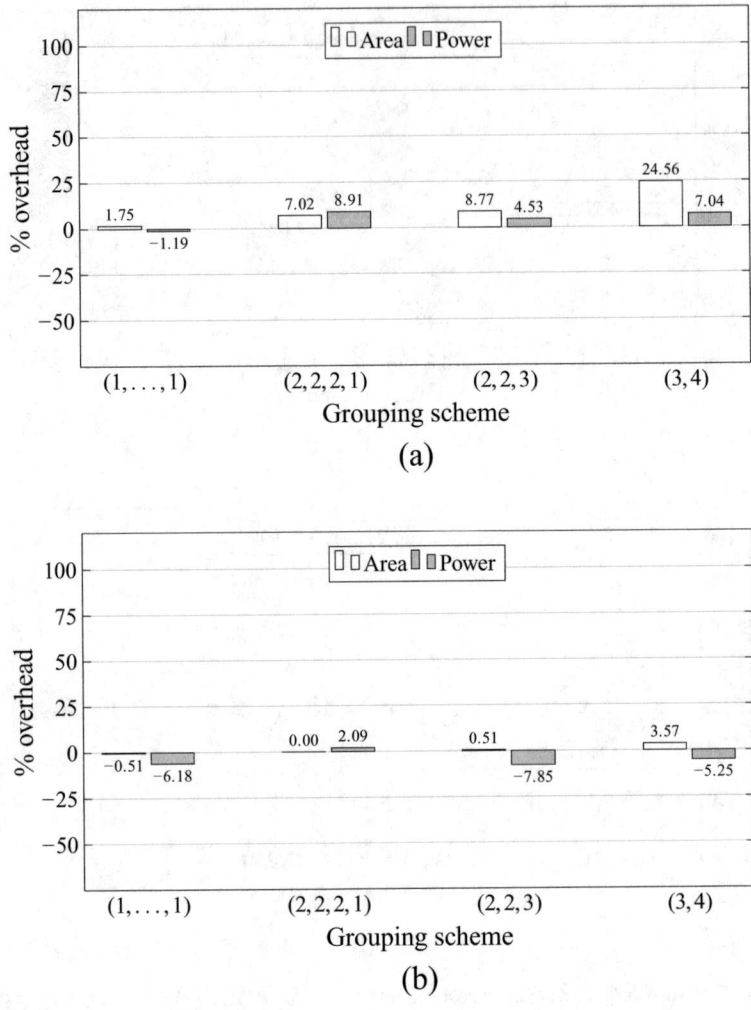

Fig. D.7 Control-Lock **low-performance** cost evaluations for the RI5CY (**a**) decoder and (**b**) ALU

input locality vectors. The output layer contains two outputs—one per key-bit value. Since the MLP model operates on a 1-dimensional input, the extracted localities are directly fed to the input layer. All other MLP parameters are shown in Table D.2a.

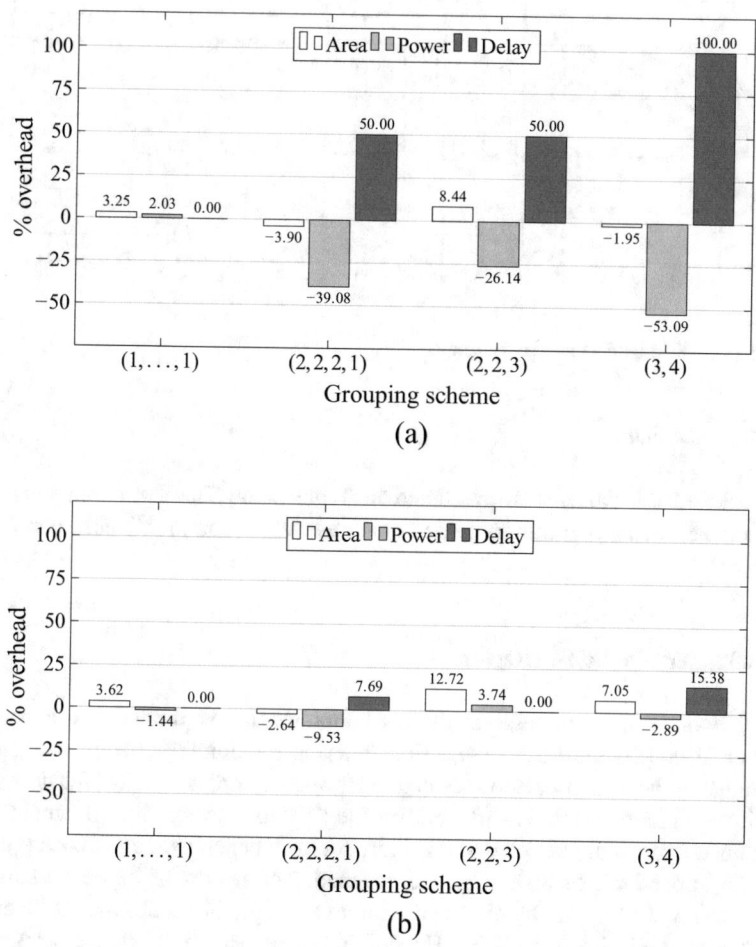

Fig. D.8 Control-Lock **high-performance** cost evaluations for the RI5CY (**a**) decoder and (**b**) ALU

D.4.2   CNN Evolution and Design Parameters

D.4.2.1   CNN Setup

As discussed in Appendix C.3, some parts of the CNN architecture and parameters are fixed. These hyperparameters were set to commonly used values from literature [181]. Dropout is not used since overfitting has not been observed in the evaluation. The maximum number of internal layers is limited to *seven*. To accommodate for the operation mode of CNNs, the locality vectors are reshaped to the dimension 20x20x1. All other CNN hyperparameters are shown in Table D.2b.

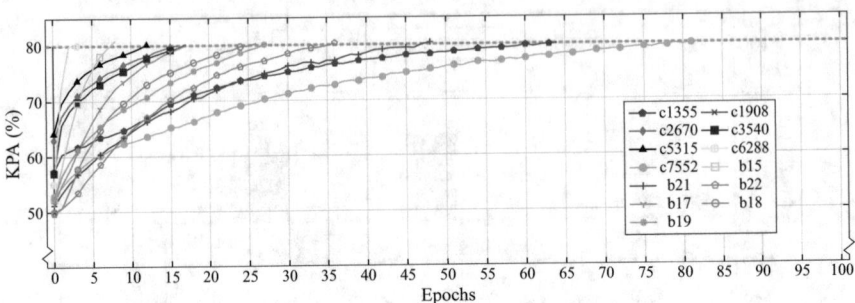

Fig. D.9 CNN training convergence graph

D.4.2.2   GA Setup

The selected GA parameters are listed in Table D.2a. The GA parameters were selected based on a preliminary tuning phase as well as commonly used parameters and operators.

D.4.2.3   Termination Criterion

In the evolutionary process, every CNN architecture is trained for a selected number of epochs until the termination criterion for training is triggered. Hence, a termination condition must be defined in advance. In order to adjust the number of epochs to an amount that is sufficient for the CNN to converge for all benchmarks, we trained a randomly selected architecture for each benchmark and attack scenario until an upper limit of 80% KPA was reached. The results of this evaluation are shown in Fig. D.9. It can be observed that, on average, 44 epochs are sufficient for the CNN to reach the target KPA. Hence, this value was selected as a termination criterion. The reasoning behind this concept is that the exploratory search of CNN architectures can make an educated decision based on a number of epochs that is sufficient to capture the learning asymptote and filter out fitter solutions. Thus, it is not necessary to train the CNN instances to their limits. Once the best network is selected, the number of epochs can be increased before deployment.

D.4.3   SnapShot: Evolved CNN Architectures

The best evolved (fittest) CNN architectures for GSS and SRS are shown in Tables D.3 and D.4, respectively. In GSS, only one final network is evolved since the training set is equivalent for all targets. In comparison, SRS results in a single network for each target netlist. The GSS network is constructed on five internal layers, whereas the SRS networks have only up to three internal layers. This suggests that the GSS

Table D.2 GA, MLP, and CNN parameters

(a) GA

Parameter	Value
GA algorithm type	Generational
Population size	10
Representation	Bitstring
# generations	20
Fitness	KPA
# epochs	44
Mutation operation	Bit-flip
Mutation probability	0.1
Crossover probability	0.9
Selection operator	Tournament (3)
Crossover operator	Two-point
Gen. of initial population	Random

(b) MLP

Parameter	Value
Layer types	Dense
Input layer activation func.	ReLU
Hidden layer activation func.	ReLU
Output layer activation func.	Softmax
GSS #units per layer	$400 \times 1000 \times 256 \times 2$
SRS #units per layer	$400 \times 512 \times 128 \times 2$
Batch size	128

(c) CNN

Parameter	Value
Input/convolutional/dense layer activation function	ReLU
Convolutional layer activation function	ReLU
Dense layer activation function	ReLU
Output layer activation function	Softmax
Loss functions	Sparse categorical cross entropy
Input/convolution layer kernel size	3×3
# filters in input/convolution layer	64
# filters in convolution layer	128
# units in dense layer	128
# units in output layer	2
Pool size	2×2
Stride size	1
Batch size	128
Optimizer	Adam
Learning rate	0.001
Beta_1	0.9
Beta_2	0.999

Table D.3 Evolved CNN architecture (genotype) for GSS

IC	[[Layer Presence], [Layer Type]]
Training Benchmarks	[[0, 1, 1, 1, 1, 0, 1], [0, 1, 1, 1, 1, 1, 0]]

Table D.4 Evolved CNN architectures (genotypes) for SRS

(a) ISCASC'85

IC	[[Layer Presence], [Layer Type]]
c1355	[[0, 1, 0, 0, 0, 0, 1], [1, 0, 1, 1, 1, 0, 1]]
c1908	[[0, 0, 0, 1, 0, 0, 0], [0, 0, 1, 1, 0, 0, 0]]
c2670	[[0, 1, 0, 0, 0, 0, 0], [0, 0, 1, 1, 1, 1, 1]]
c3540	[[0, 0, 0, 0, 1, 1, 0], [1, 0, 0, 0, 1, 0, 0]]
c5315	[[0, 0, 0, 0, 1, 1, 0], [1, 0, 0, 0, 1, 0, 0]]
c6288	[[0, 0, 0, 0, 1, 1, 0], [1, 0, 0, 0, 1, 0, 0]]
c7552	[[1, 1, 1, 0, 0, 0, 0], [0, 1, 0, 1, 0, 0, 1]]

(b) ITC'99

IC	[[Layer Presence], [Layer Type]]
b15	[[0, 0, 0, 0, 0, 0, 1], [1, 0, 0, 0, 1, 1, 1]]
b21	[[0, 1, 0, 0, 0, 0, 0], [1, 1, 1, 0, 1, 0, 0]]
b22	[[0, 0, 0, 0, 0, 0, 1,], [1, 1, 1, 0, 0, 0, 1]]
b17	[[0, 0, 1, 0, 0, 0, 0], [1, 1, 1, 0, 0, 0, 0]]
b18	[[0, 0, 0, 0, 0, 0, 1], [0, 0, 0, 0, 0, 1, 1]]
b19	[[0, 0, 1, 0, 0, 0, 0], [1, 0, 1, 0, 0, 1, 1]]

training set is more versatile, hence requiring more expressive CNN architectures to perform the feature extraction. Moreover, the low termination criterion of 44 epochs together with the low number of internal layers indicate that the training process easily converges. Consequently, these observations suggest that the classification of XOR- and XNOR-driven localities is not a particularly difficult problem—hence, the security foundation of these locking policies must be further revised.

D.5   D-MUX: Resilience Evaluation

The following details the experimental setup that was used to evaluate the Deceptive Multiplexer Logic Locking (D-MUX) scheme against the SWEEP [4] and Snap-Shot [169] attack, as discussed in Chap. 9.

D.5.1   SWEEP: Attack Setup

Metrics   As defined in [4], the evaluation of SWEEP utilizes two metrics: *accuracy* and *precision*. Accuracy is defined as the percentage of correctly extracted key-bit values out of the entire key, *regardless of potential X values*:

$$Accuracy = \frac{N_{correct}}{N_{total}} \cdot 100\% \qquad (D.1)$$

Precision is defined as the percentage of correctly guessed keys, where every potential X value is regarded as a correct guess:

$$Precision = \frac{N_{correct} + N_X}{N_{total}} \cdot 100\% \qquad (D.2)$$

Furthermore, the attacker can set the margin m to control the freedom of the attack to make "wild" guesses when SWEEP is not able to conclusively decide the correct key-bit value. By default, $m = 0$.

Experimental Setup The SWEEP tool provided by [4] was used for the attack evaluation. The setup is as follows. First, all benchmarks from Table 9.2a, Chap. 9, were copied 100 times and locked with random 64-bit keys. Thus, a data set of 700 locked benchmarks was generated. Second, for each benchmark, the data set was divided into a test set and a training set. The test set consists of 100 locked target netlists. The training set consists of the remaining 600 other benchmarks. The attack is repeated for a range of margins ($m \in [0.0, 0.01]$) with a step of size $6.25 \cdot 10^{-3}$.

D.5.2   SnapShot: Attack Setup and Evaluation

Experimental Setup Both D-MUX variants were evaluated for both GSS and SRS. For GSS, the training set is constructed using all ISCAS'85 and ITC'99 benchmarks from Table 9.2. Each benchmark was copied and locked 1,000 times using randomly generated 64-bit keys, resulting in 13,000 locked netlists and 832,000 labeled localities. The test set consisted of all benchmarks from Table 9.2c, where each netlist was locked 1000 times, yielding 448,000 unlabeled localities. For SRS, the training set was generated by relocking each target netlist 1000 times to predict the key of the initial target. Hereby, all ISCAS'85 and ITC'99 benchmarks from Table 9.2 were selected for locking. This process was repeated 20 times to get an average value for all benchmarks.

Detailed Results The GSS results of the SnapShot evaluation on D-MUX for the MLP and CNN model are presented in Fig. D.12 for each benchmark individually. The SRS results of the SnapShot evaluation on D-MUX are presented in Fig. D.13 for each benchmark set individually.

D.5.3   *Localities Extraction for MUX-Based Locking*

An example of the locality extraction for a 1-bit key input for Multiplexer (MUX)-based logic locking is presented in Fig. D.10. The extraction parameters are equivalent to the example in Sect. 9.2.2.2. The extraction is adjusted to encode MUXs by capturing all gates that lead to or from a selected set of key-controlled multiplexers, as visualized in Fig. D.11. To simplify the representation, only one encoding is stored in the case of two inserted multiplexers. The extracted locality vector consists of the following components.

- $K_{act}[i]$: the key-bit value (in case the vector is labeled).
- l_b: the backward path; the input cones of f_i and f_j.
- l_{kg}: the key gate in the form of one MUX gate (regardless of whether one or two MUX gates are inserted).
- l_{sb}^i: the backward path specific to g_1^i.
- l_{sb}^j: the backward path specific to g_1^j (if not existent, l_{sb}^j is filled with zeros).
- l_f: the forward path; the output cones of g_i and g_j.

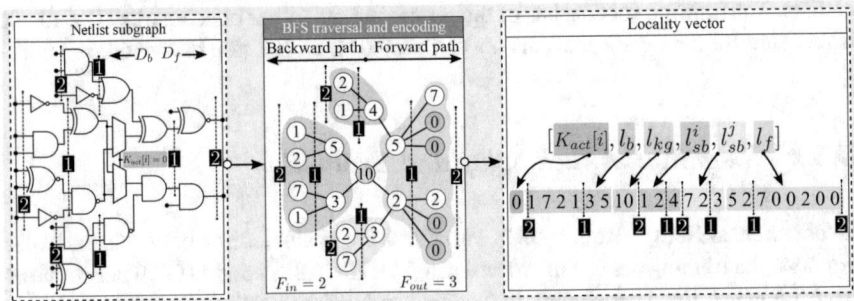

Fig. D.10 Example of localities extraction for MUX-based locking

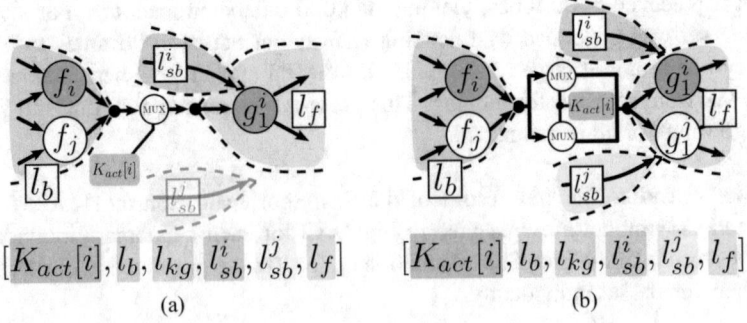

Fig. D.11 Locality vectors for D-MUX. (a) S_1 to S_3. (b) S_4

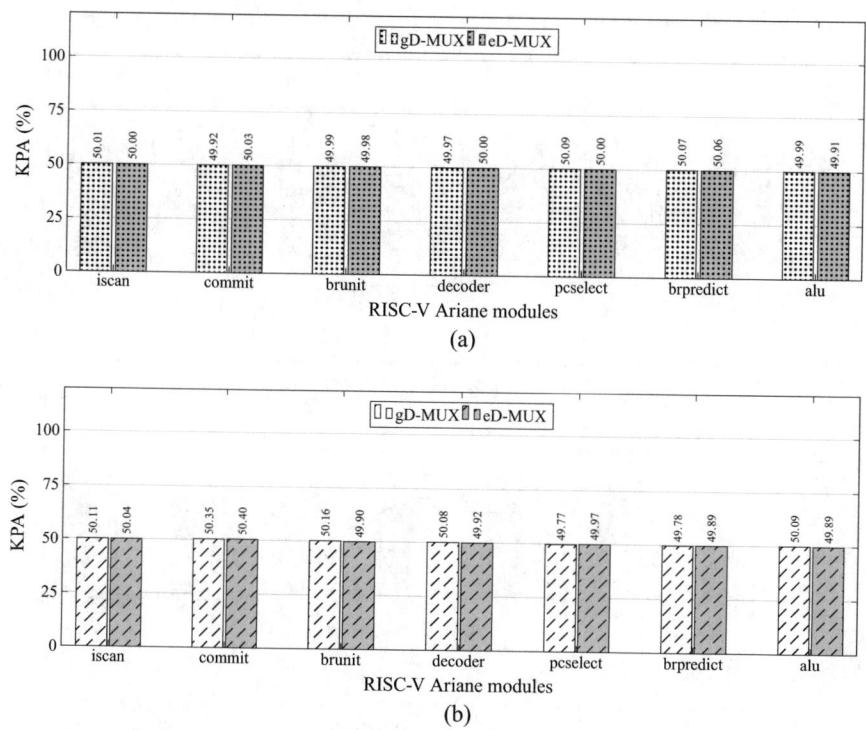

Fig. D.12 SnapShot attack evaluation on D-MUX for the generalized set scenario. (**a**) MLP for GSS. (**b**) CNN for GSS

D.6   D-MUX Cost Evaluation

To offer a detailed view of the cost implications of D-MUX, the selected ISCAS'85 and ITC'99 benchmarks (see Chap. 9) were locked with 64-bit keys using Generalized D-MUX (gD-MUX) and Enhanced D-MUX (eD-MUX) separately. Each benchmark is synthesized for a range of clock periods (T_{clk}) to generate a reasonable post-synthesis AT plot (see Appendix D.2). The AT plots for the ISCAS'85 and ITC'99 benchmark are shown in Figs. D.14, D.15, D.16, D.17, and D.18. Each point in the plots presents an average value across 20 synthesis rounds for 20 locked netlist instances to include locking and synthesis variations. Furthermore, a detailed comparison of the average area, power, and delay overhead for each benchmark separately is presented in Figs. D.19, D.20, and D.21 for AT-optimum points, Figs. D.22 and D.23 for low-performance points, and Figs. D.24, D.25, and D.26 for high-performance points.

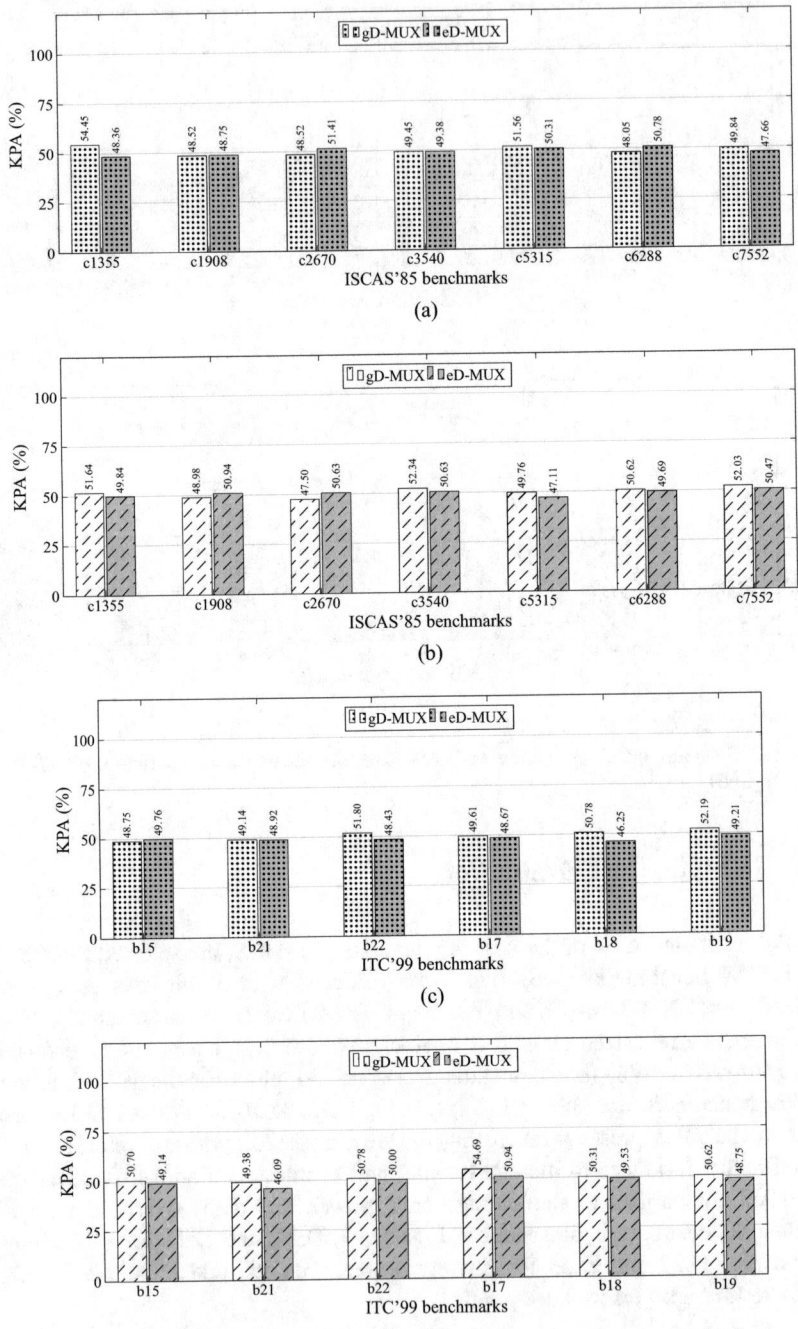

Fig. D.13 SnapShot attack evaluation on D-MUX for the self-referencing scenario. (**a**) MLP for SRS on ISCAS'85 benchmarks. (**b**) CNN for SRS on ISCAS'85 benchmarks. (**c**) MLP for SRS on ITC'99 benchmarks. (**d**) CNN for SRS on ITC'99 benchmarks

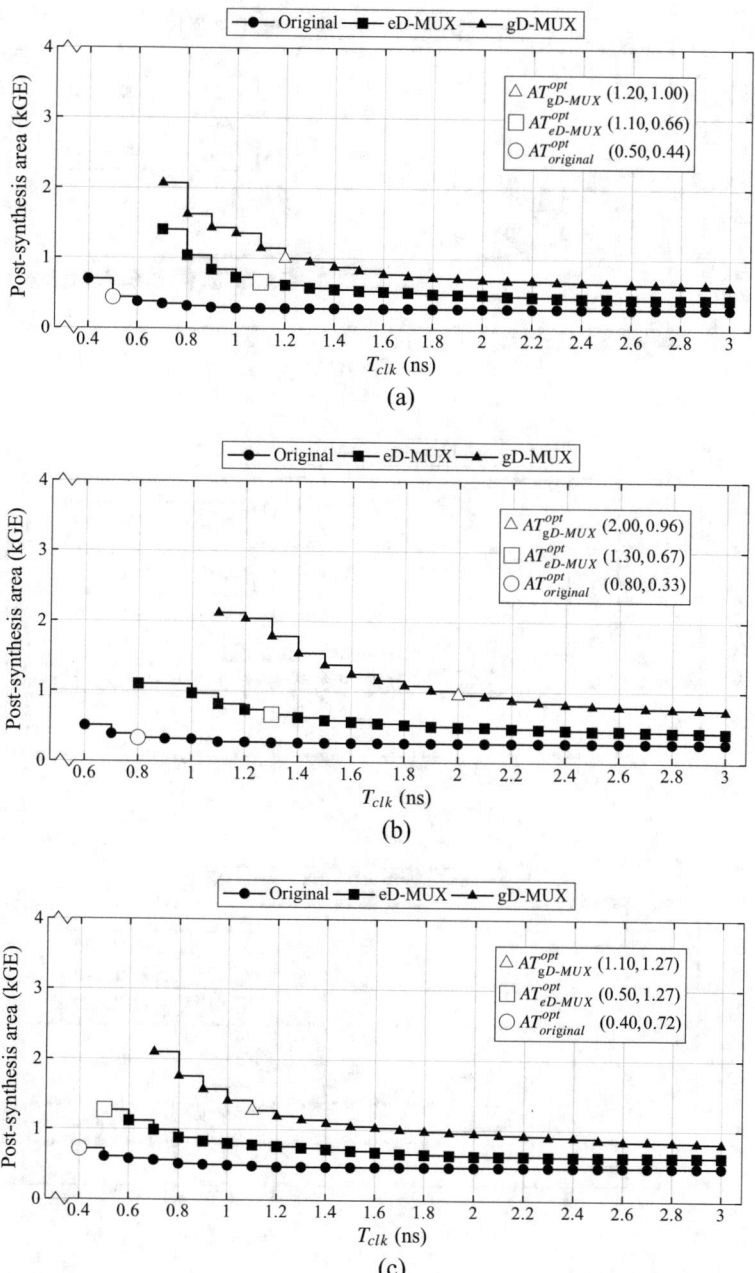

Fig. D.14 D-MUX AT evaluation for ISCAS'85 benchmarks (1). (**a**) c1355. (**b**) c1908. (**c**) c2670

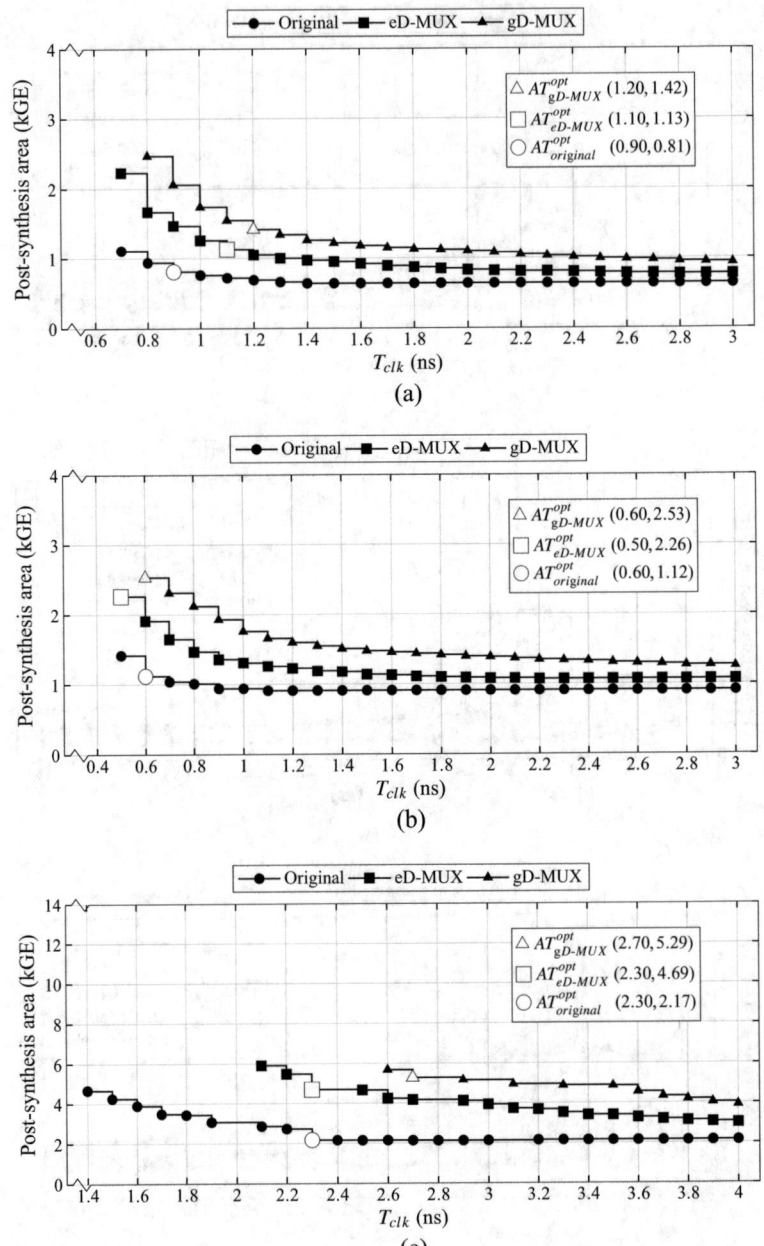

Fig. D.15 D-MUX AT evaluation for ISCAS'85 benchmarks (2). (**a**) c3540. (**b**) c5315. (**c**) c6288

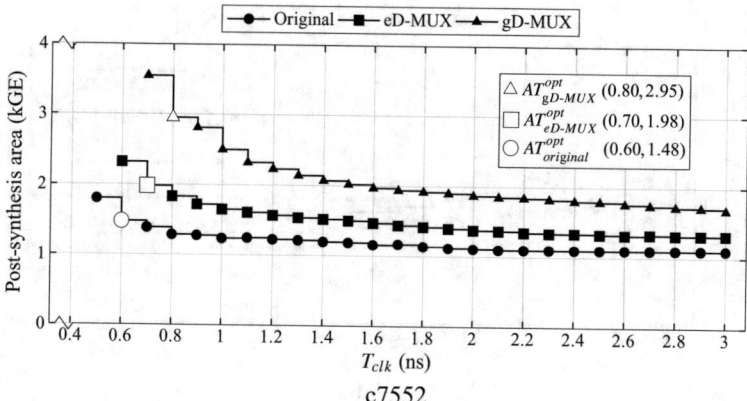

c7552

Fig. D.16 D-MUX AT evaluation for ISCAS'85 benchmarks (3)

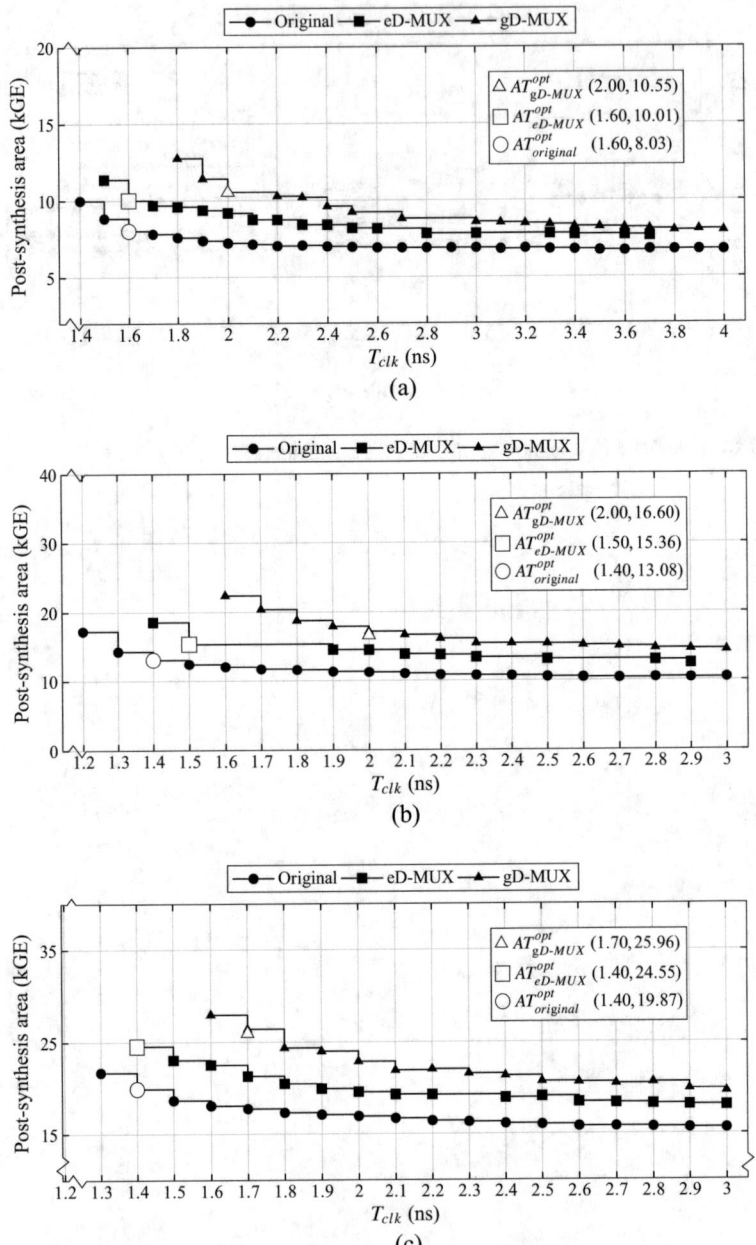

Fig. D.17 D-MUX AT evaluation for ITC'99 benchmarks (1). (**a**) b15. (**b**) b21. (**c**) b22

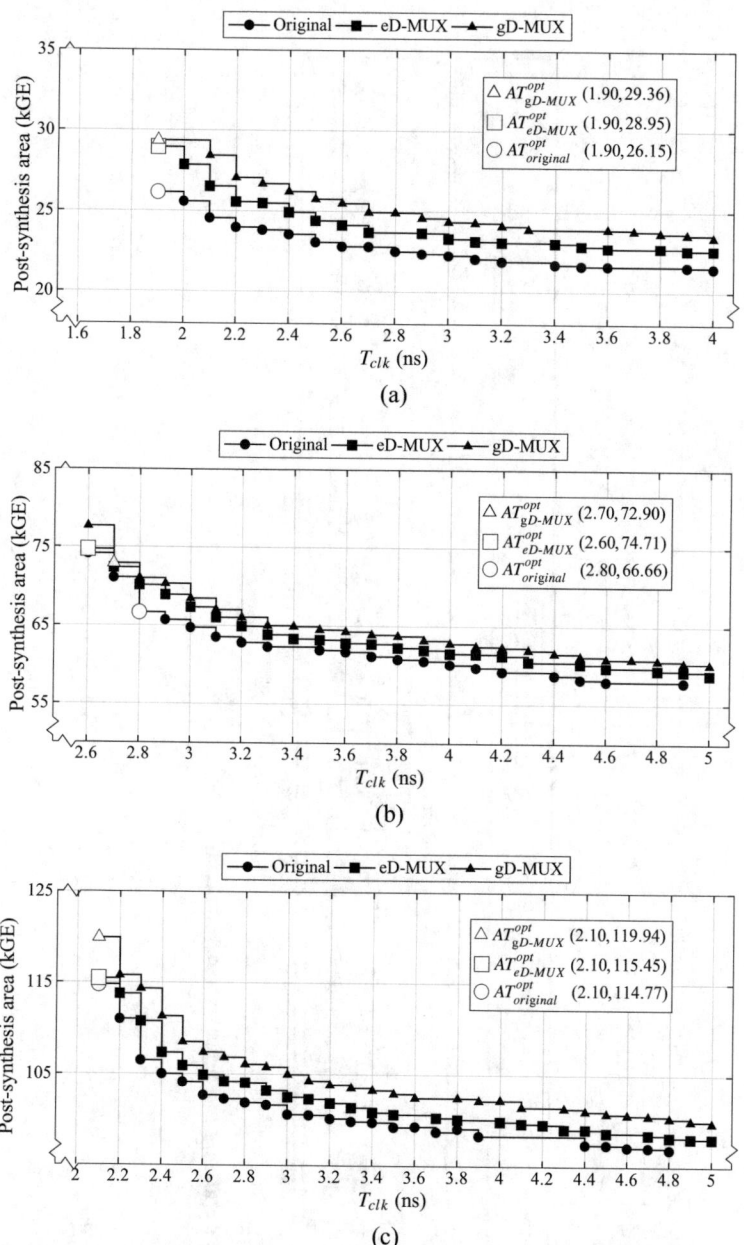

Fig. D.18 D-MUX AT evaluation for ITC'99 benchmarks (2). (**a**) b17. (**b**) b18. (**c**) b19

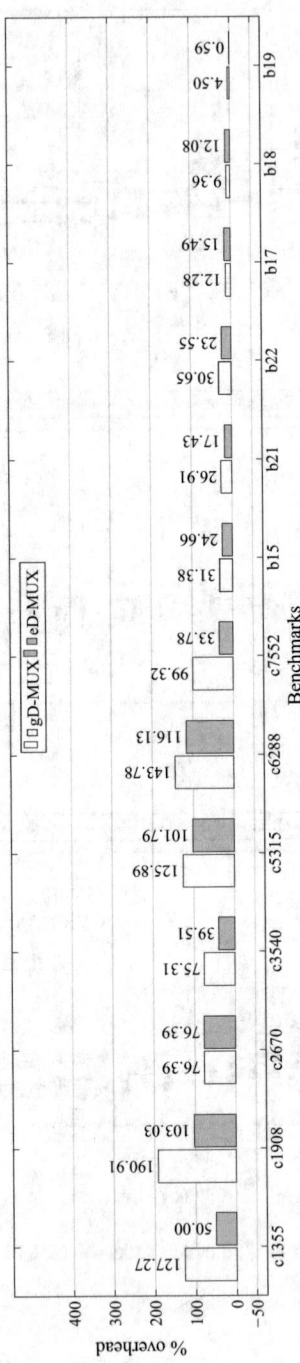

Fig. D.19 D-MUX **AT-optimum** cost evaluation (area overhead)

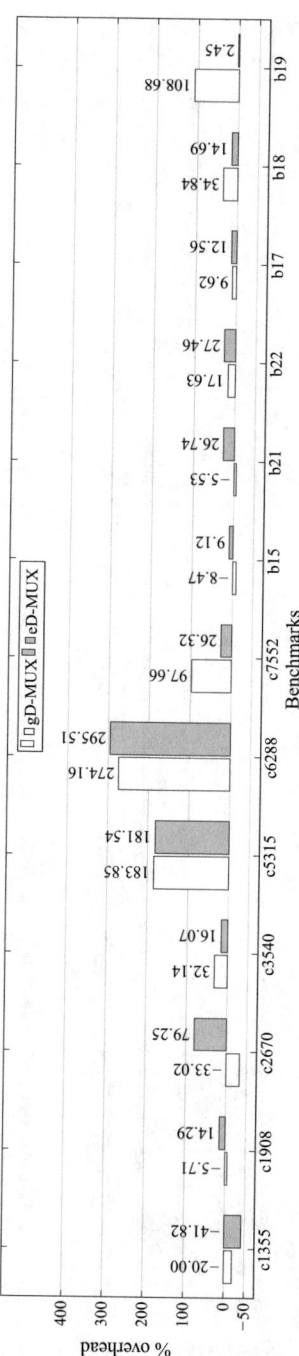

Fig. D.20 D-MUX **AT-optimum** cost evaluation (power overhead)

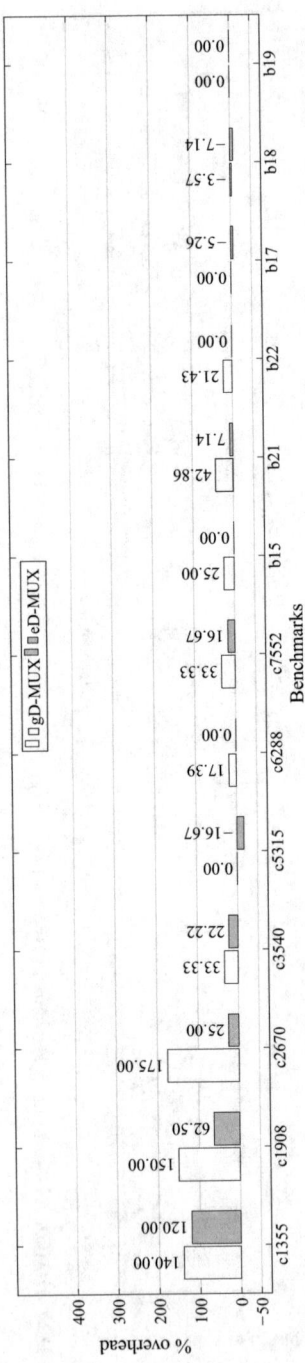

Fig. D.21 D-MUX **AT-optimum** cost evaluation (delay overhead)

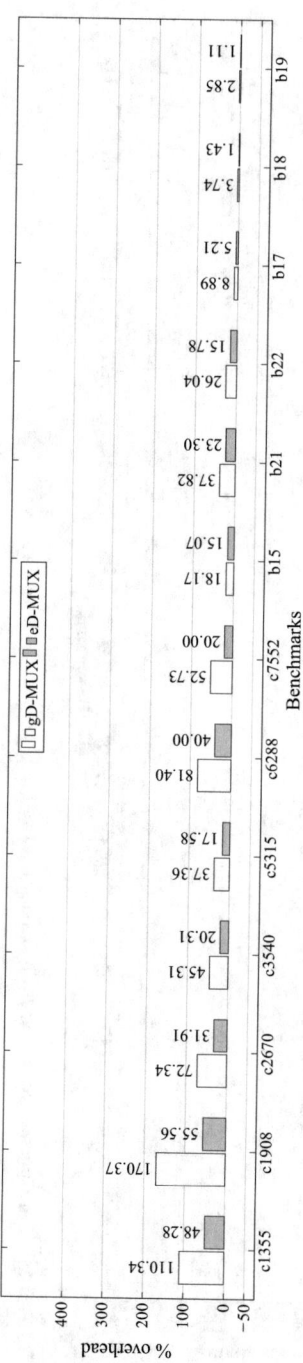

Fig. D.22 D-MUX low-performance cost evaluation (area overhead)

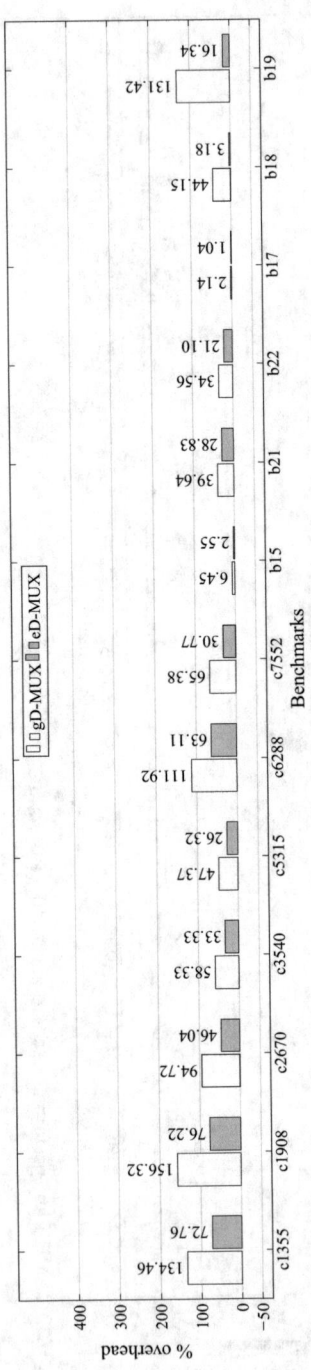

Fig. D.23 D-MUX low-performance cost evaluation (power overhead)

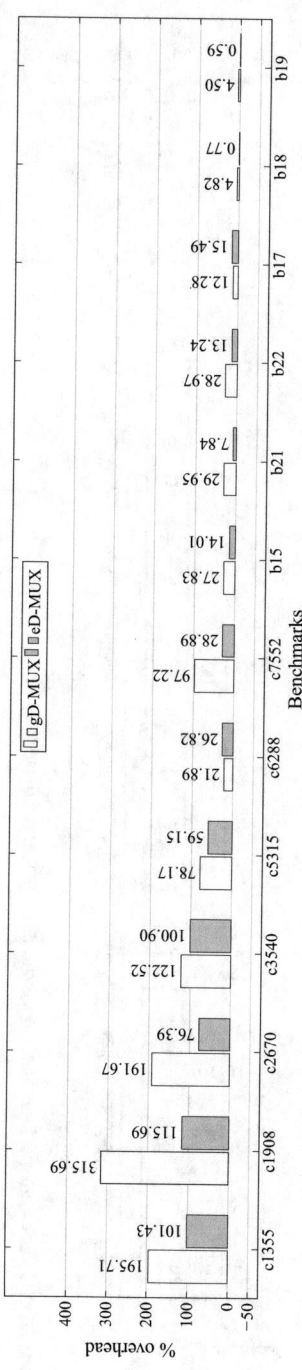

Fig. D.24 D-MUX **high-performance** cost evaluation (area overhead)

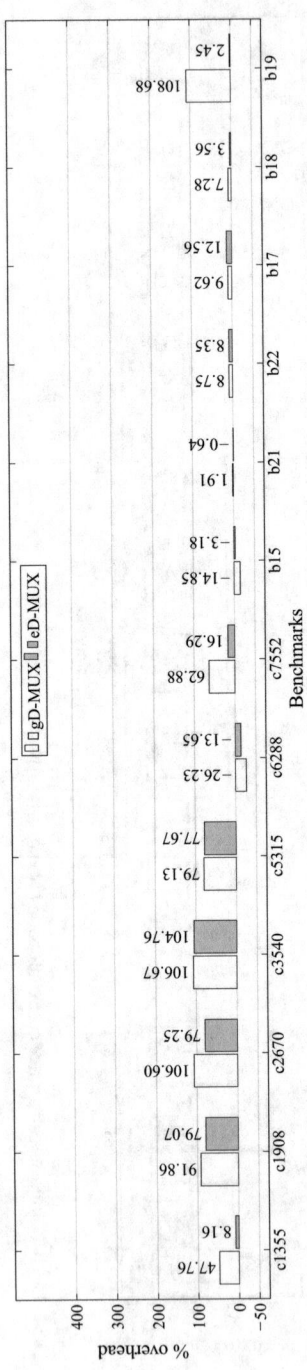

Fig. D.25 D-MUX high-performance cost evaluation (power overhead)

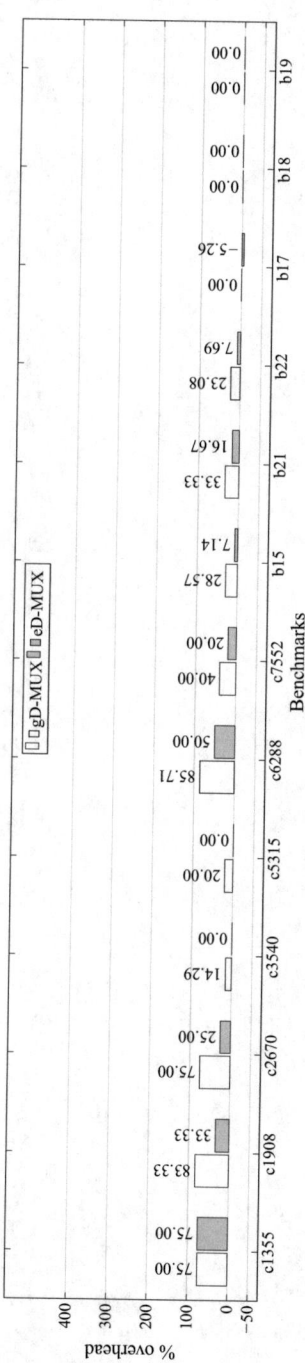

Fig. D.26 D-MUX **high-performance** cost evaluation (delay overhead)

References

1. M. Abadi, A. Agarwal, P. Barham, E. Brevdo, Z. Chen, C. Citro, G.S. Corrado, A. Davis, J. Dean, M. Devin, S. Ghemawat, I. Goodfellow, A. Harp, G. Irving, M. Isard, Y. Jia, R. Jozefowicz, L. Kaiser, M. Kudlur, J. Levenberg, D. Mané, R. Monga, S. Moore, D. Murray, C. Olah, M. Schuster, J. Shlens, B. Steiner, I. Sutskever, K. Talwar, P. Tucker, V. Vanhoucke, V. Vasudevan, F. Viégas, O. Vinyals, P. Warden, M. Wattenberg, M. Wicke, Y. Yu, X. Zheng, TensorFlow: large-scale machine learning on heterogeneous systems (2015). https://www.tensorflow.org/. Software available from tensorflow.org
2. S. Adee, The hunt for the kill switch. IEEE Spectr. **45**(5), 34–39 (2008). https://doi.org/10.1109/MSPEC.2008.4505310
3. A. Alaql, S. Bhunia, Scalable attack-resistant obfuscation of logic circuits (2020). https://arxiv.org/abs/2010.15329
4. A. Alaql, D. Forte, S. Bhunia, Sweep to the secret: a constant propagation attack on logic locking, in *2019 Asian Hardware Oriented Security and Trust Symposium (AsianHOST)* (2019), pp. 1–6. https://doi.org/10.1109/AsianHOST47458.2019.9006720
5. A. Alaql, M.M. Rahman, S. Bhunia, Scope: synthesis-based constant propagation attack on logic locking. IEEE Trans. Very Large Scale Integr. Syst. **29**(8), 1529–1542 (2021). https://doi.org/10.1109/TVLSI.2021.3089555
6. Y.M. Alkabani, F. Koushanfar, Active hardware metering for intellectual property protection and security, in *16th USENIX Security Symposium (USENIX Security 07)* (USENIX Association, Boston, MA, 2007). https://www.usenix.org/conference/16th-usenix-security-symposium/active-hardware-metering-intellectual-property-protection
7. L.M. Almeida, T.B. Ludermir, An evolutionary approach for tuning artificial neural network parameters, in: *HAIS*, ed. by E. Corchado, A. Abraham, W. Pedrycz (Springer Berlin Heidelberg, Berlin, Heidelberg, 2008), pp. 156–163
8. L. Alrahis, J. Knechtel, F. Klemme, H. Amrouch, O. Sinanoglu, Gnn4rel: graph neural networks for predicting circuit reliability degradation (2022). https://doi.org/10.48550/ARXIV.2208.02868
9. L. Alrahis, S. Patnaik, M.A. Hanif, M. Shafique, O. Sinanoglu, Untangle: unlocking routing and logic obfuscation using graph neural networks-based link prediction, in *2021 IEEE/ACM International Conference On Computer Aided Design (ICCAD)* (2021), pp. 1–9. https://doi.org/10.1109/ICCAD51958.2021.9643476
10. L. Alrahis, S. Patnaik, F. Khalid, M.A. Hanif, H. Saleh, M. Shafique, O. Sinanoglu, GNNUnlock: graph neural networks-based oracle-less unlocking scheme for provably secure logic locking. Accepted at DATE 2021 (2020). https://arxiv.org/abs/2012.05948

© The Author(s), under exclusive license to Springer Nature Switzerland AG 2023
D. Sisejkovic, R. Leupers, *Logic Locking*,
https://doi.org/10.1007/978-3-031-19123-7

11. L. Alrahis, S. Patnaik, J. Knechtel, H. Saleh, B. Mohammad, M. Al-Qutayri, O. Sinanoglu, UNSAIL: thwarting oracle-less machine learning attacks on logic locking. IEEE Trans. Inform. Forensics Secur. **16**, 2508–2523 (2021). https://doi.org/10.1109/TIFS.2021.3057576

12. L. Alrahis, S. Patnaik, M. Shafique, O. Sinanoglu, Circumventing learning-resilient MUX-locking using graph neural network-based link prediction (2021). https://doi.org/10.48550/ARXIV.2112.07178

13. L. Alrahis, S. Patnaik, M. Shafique, O. Sinanoglu, OMLA: an oracle-less machine learning-based attack on logic locking. IEEE Trans. Circuits Syst. II: Exp. Briefs **69**(3), 1602–1606 (2022). https://doi.org/10.1109/TCSII.2021.3113035

14. L. Alrahis, M. Yasin, H. Saleh, B. Mohammad, M. Al-Qutayri, Functional reverse engineering on SAT-attack resilient logic locking, in *2019 IEEE International Symposium on Circuits and Systems (ISCAS)* (2019), pp. 1–5. https://doi.org/10.1109/ISCAS.2019.8702704

15. S. Amir, B. Shakya, D. Forte, M. Tehranipoor, S. Bhunia, Comparative analysis of hardware obfuscation for IP protection, in *Proceedings of the on Great Lakes Symposium on VLSI 2017, GLSVLSI '17* (Association for Computing Machinery, New York, 2017), pp. 363–368. https://doi.org/10.1145/3060403.3060495

16. S. Amir, B. Shakya, X. Xu, Y. Jin, S. Bhunia, M. Tehranipoor, D. Forte, Development and evaluation of hardware obfuscation benchmarks. J. Hardw. Syst. Secur. **2**(2), 142–161 (2018). https://doi.org/10.1007/s41635-018-0036-3

17. M.A. Arbib (ed.), *The Handbook of Brain Theory and Neural Networks* (MIT Press, Cambridge, MA, 1998)

18. K.Z. Azar, H.M. Kamali, H. Homayoun, A. Sasan, SMT attack: next generation attack on obfuscated circuits with capabilities and performance beyond the SAT attacks. IACR Trans. Cryptogr. Hardw. Embedded Syst. **2019**(1), 97–122 (2018). https://doi.org/10.13154/tches.v2019.i1.97-122

19. K.Z. Azar, H.M. Kamali, H. Homayoun, A. Sasan, NNgSAT: neural network guided SAT attack on logic locked complex structures, in *Proceedings of the 39th International Conference on Computer-Aided Design, ICCAD '20* (Association for Computing Machinery, New York, 2020). https://doi.org/10.1145/3400302.3415669

20. L. Azriel, R. Ginosar, A. Mendelson, Sok: an overview of algorithmic methods in IC reverse engineering, in *Proceedings of the 3rd ACM Workshop on Attacks and Solutions in Hardware Security Workshop, ASHES'19* (Association for Computing Machinery, New York, 2019), pp. 65–74. https://doi.org/10.1145/3338508.3359575

21. P. Ba, S. Dupuis, M. Palanichamy, M. Flottes, G. Di Natale, B. Rouzeyre, Hardware trust through layout filling: a hardware Trojan prevention technique, in *2016 IEEE Computer Society Annual Symposium on VLSI (ISVLSI)* (2016), pp. 254–259. https://doi.org/10.1109/ISVLSI.2016.22

22. J. Baehr, A. Bernardini, G. Sigl, U. Schlichtmann, Machine learning and structural characteristics for reverse engineering, in *Proceedings of the 24th Asia and South Pacific Design Automation Conference, ASPDAC '19* (Association for Computing Machinery, New York, 2019), pp. 96–103. https://doi.org/10.1145/3287624.3288740

23. J. Baehr, A. Hepp, M. Brunner, M. Malenko, G. Sigl, Open source hardware design and hardware reverse engineering: a security analysis. EasyChair Preprint no. 8604 (EasyChair, 2022)

24. A. Baumgarten, A. Tyagi, J. Zambreno, Preventing IC piracy using reconfigurable logic barriers. IEEE Design Test Comput. **27**(1), 66–75 (2010). https://doi.org/10.1109/MDT.2010.24

25. S. Becker, C. Wiesen, N. Albartus, N. Rummel, C. Paar, An exploratory study of hardware reverse engineering—technical and cognitive processes, in *Sixteenth Symposium on Usable Privacy and Security (SOUPS 2020)* (USENIX Association, 2020), pp. 285–300. https://www.usenix.org/conference/soups2020/presentation/becker

26. S. Bhattacharjee, J. Tang, M. Ibrahim, K. Chakrabarty, R. Karri, Locking of biochemical assays for digital microfluidic biochips, in *2018 IEEE 23rd European Test Symposium (ETS)* (2018), pp. 1–6. https://doi.org/10.1109/ETS.2018.8400686

27. S. Bhattacharjee, J. Tang, S. Poddar, M. Ibrahim, R. Karri, K. Chakrabarty, Bio-chemical assay locking to thwart bio-IP theft. ACM Trans. Des. Autom. Electron. Syst. **25**(1) (2019). https://doi.org/10.1145/3365579

28. S. Bhunia, M.S. Hsiao, M. Banga, S. Narasimhan, Hardware trojan attacks: threat analysis and countermeasures. Proc. IEEE **102**(8), 1229–1247 (2014). https://doi.org/10.1109/JPROC. 2014.2334493

29. S. Bhunia, M. Tehranipoor, *Hardware Security: A Hands-on Learning Approach*, 1st edn. (Morgan Kaufmann Publishers, San Francisco, 2018)

30. S. Bhunia, M. Tehranipoor, *The Hardware Trojan War: Attacks, Myths, and Defenses* (Springer, 2018). https://doi.org/10.1007/978-3-319-68511-3

31. F. Brglez, D. Bryan, K. Kozminski, Combinational profiles of sequential benchmark circuits, in *IEEE International Symposium on Circuits and Systems*, vol. 3 (1989), pp. 1929–1934. https://doi.org/10.1109/ISCAS.1989.100747

32. Bundesministerium für Bildung und Forschung (BMBF), Mikroelektronik. Vertrauenswürdig und nachhaltig. Für Deutschland und Europa. Rahmenprogramm der Bundesregierung für Forschung und Innovation 2021–2024 (2021). https://www.elektronikforschung.de/ rahmenprogramm

33. Bundesministerium für Bildung und Forschung (BMBF), Vertrauenswürdige Elektronik. Forschung und Innovation für technologische Souveränität (2021). https://www. elektronikforschung.de/service/publikationen/vertrauenswuerdige-elektronik

34. A. Chakraborty, N.G. Jayasankaran, Y. Liu, J. Rajendran, O. Sinanoglu, A. Srivastava, Y. Xie, M. Yasin, M. Zuzak, Keynote: a disquisition on logic locking. IEEE Trans. Comput.-Aided Design Integr. Circuits Syst. **39**(10), 1952–1972 (2020). https://doi.org/10.1109/TCAD.2019. 2944586

35. A. Chakraborty, Y. Liu, A. Srivastava, Evaluating the security of delay-locked circuits. IEEE Trans. Comput.-Aided Design Integr. Circuits Syst. **40**(4), 608–619 (2021). https://doi.org/ 10.1109/TCAD.2020.3008843

36. P. Chakraborty, J. Cruz, S. Bhunia, SAIL: machine learning guided structural analysis attack on hardware obfuscation, in *2018 Asian Hardware Oriented Security and Trust Symposium (AsianHOST)* (2018), pp. 56–61. https://doi.org/10.1109/AsianHOST.2018.8607163

37. P. Chakraborty, J. Cruz, S. Bhunia, SURF: joint structural functional attack on logic locking, in *2019 IEEE International Symposium on Hardware Oriented Security and Trust (HOST)* (2019), pp. 181–190. https://doi.org/10.1109/HST.2019.8741028

38. R.S. Chakraborty, S. Bhunia, Harpoon: an obfuscation-based SOC design methodology for hardware protection. IEEE Trans. Comput.-Aided Design Integr. Circuits Syst. **28**(10), 1493–1502 (2009). https://doi.org/10.1109/TCAD.2009.2028166

39. R.S. Chakraborty, S. Narasimhan, S. Bhunia, Hardware Trojan: threats and emerging solutions, in *2009 IEEE International High Level Design Validation and Test Workshop* (2009), pp. 166–171. https://doi.org/10.1109/HLDVT.2009.5340158

40. H. Chen, C. Fu, J. Zhao, F. Koushanfar, GenUnlock: an automated genetic algorithm framework for unlocking logic encryption, in *2019 IEEE/ACM International Conference on Computer-Aided Design (ICCAD)* (2019), pp. 1–8. https://doi.org/10.1109/ICCAD45719. 2019.8942134

41. H.Y. Chiang, Y.C. Chen, D.X. Ji, X.M. Yang, C.C. Lin, C.Y. Wang, LOOPLock: logic optimization-based cyclic logic locking. IEEE Trans. Comput.-Aided Design Integr. Circuits Syst. **39**(10), 2178–2191 (2020). https://doi.org/10.1109/TCAD.2019.2960351

42. L.W. Chow, J.P. Baukus, B.J. Wang, R.P. Cocchi, Camouflaging a standard cell based integrated circuit. U.S. Patent no 8151235B25, 2012

43. G.K. Contreras, M.T. Rahman, M. Tehranipoor, Secure split-test for preventing IC piracy by untrusted foundry and assembly, in *2013 IEEE International Symposium on Defect and Fault Tolerance in VLSI and Nanotechnology Systems (DFTS)* (2013), pp. 196–203. https://doi.org/ 10.1109/DFT.2013.6653606

44. F. Corno, M.S. Reorda, G. Squillero, RT-level ITC'99 benchmarks and first ATPG results. IEEE Design Test Comput. **17**(3), 44–53 (2000). https://doi.org/10.1109/54.867894

45. J. Daemen, V. Rijmen, *The Design of Rijndael*, vol. 2 (Springer, Berlin, 2002)
46. Defense Advanced Research Project Agency (DARP), Integrity and reliability of integrated circuits (IRIS) (2021). https://www.darpa.mil/program/integrity-and-reliability-of-integrated-circuits
47. Defense Advanced Research Project Agency (DARP), Supply chain hardware integrity for electronics defense (SHIELD) (2021). https://www.darpa.mil/program/supply-chain-hardware-integrity-for-electronics-defense
48. Defense Advanced Research Project Agency (DARP), Trusted integrated circuits (TRUST) (2021). https://www.darpa.mil/program/trusted-integrated-circuits
49. Defense Science Board Task Force, *High Performance Microchip Supply*. Annual Report. Defense Technical Information Center (DTIC), USA (2005). https://apps.dtic.mil/sti/citations/ADA435563
50. J. Dofe, Q. Yu, Novel dynamic state-deflection method for gate-level design obfuscation. IEEE Trans. Comput.-Aided Design Integr. Circuits Syst. **37**(2), 273–285 (2018). https://doi.org/10.1109/TCAD.2017.2697960
51. C. Dong, Y. Xu, X. Liu, F. Zhang, G. He, Y. Chen, Hardware Trojans in chips: a survey for detection and prevention. Sensors (Basel, Switzerland) **20**(18), 5165 (2020). https://doi.org/10.3390/s20185165
52. C. Drake, Python electronic design automation. https://pyeda.readthedocs.io/en/latest/index.html
53. S. Dupuis, P. Ba, G. Di Natale, M. Flottes, B. Rouzeyre, A novel hardware logic encryption technique for thwarting illegal overproduction and hardware trojans, in *2014 IEEE 20th International On-Line Testing Symposium (IOLTS)* (2014), pp. 49–54. https://doi.org/10.1109/IOLTS.2014.6873671
54. S. Dupuis, M.L. Flottes, Logic locking: a survey of proposed methods and evaluation metrics. J. Electron. Testing **35**(3), 273–291 (2019). https://doi.org/10.1007/s10836-019-05800-4
55. D. Duvalsaint, X. Jin, B. Niewenhuis, R.D. Blanton, Characterization of locked combinational circuits via ATPG, in *2019 IEEE International Test Conference (ITC)* (2019), pp. 1–10. https://doi.org/10.1109/ITC44170.2019.9000130
56. D. Duvalsaint, Z. Liu, A. Ravikumar, R.D. Blanton, Characterization of locked sequential circuits via ATPG, in *2019 IEEE International Test Conference in Asia (ITC-Asia)* (2019), pp. 97–102. https://doi.org/10.1109/ITC-Asia.2019.00030
57. A.E. Eiben, J.E. Smith, *Introduction to Evolutionary Computing*, 2nd edn. (Springer, Berlim, 2015)
58. R. Elnaggar, J. Chaudhuri, R. Karri, K. Chakrabarty, Learning malicious circuits in FPGA bitstreams. IEEE Trans. Comput.-Aided Design Integr. Circuits Syst. (2022). https://doi.org/10.1109/TCAD.2022.3190771
59. T. Elsken, J.H. Metzen, F. Hutter, Neural architecture search: a survey (2019). https://arxiv.org/abs/1808.05377
60. S. Engels, M. Hoffmann, C. Paar, The end of logic locking? A critical view on the security of logic locking. Cryptology ePrint Archive, Report 2019/796 (2019). https://eprint.iacr.org/2019/796
61. B. Erbagci, C. Erbagci, N.E.C. Akkaya, K. Mai, A secure camouflaged threshold voltage defined logic family, in *2016 IEEE International Symposium on Hardware Oriented Security and Trust (HOST)* (2016), pp. 229–235. https://doi.org/10.1109/HST.2016.7495587
62. A. Ferraiuolo, R. Xu, D. Zhang, A.C. Myers, G.E. Suh, Verification of a practical hardware security architecture through static information flow analysis, in *Proceedings of the Twenty-Second International Conference on Architectural Support for Programming Languages and Operating Systems, ASPLOS '17* (Association for Computing Machinery, New York, 2017), pp. 555–568. https://doi.org/10.1145/3037697.3037739
63. M. Fyrbiak, Constructive and destructive reverse engineering aspects of digital systems. Doctoral Thesis, Ruhr-Universität Bochum, Universitätsbibliothek (2019). https://doi.org/10.13154/294-6473

64. U. Guin, D. Dimase, M. Tehranipoor, Counterfeit integrated circuits: detection, avoidance, and the challenges ahead. J. Electron. Test. **30**(1), 9–23 (2014). https://doi.org/10.1007/s10836-013-5430-8

65. U. Guin, K. Huang, D. DiMase, J.M. Carulli, M. Tehranipoor, Y. Makris, Counterfeit integrated circuits: a rising threat in the global semiconductor supply chain. Proc. IEEE **102**(8), 1207–1228 (2014). https://doi.org/10.1109/JPROC.2014.2332291

66. U. Guin, Q. Shi, D. Forte, M.M. Tehranipoor, FORTIS: a comprehensive solution for establishing forward trust for protecting IPs and ICs. ACM Trans. Des. Autom. Electron. Syst. **21**(4) (2016). https://doi.org/10.1145/2893183

67. X. He, K. Zhao, X. Chu, AutoML: a survey of the state-of-the-art. Knowl.-Based Syst. **212**, 106622 (2021). https://doi.org/10.1016/j.knosys.2020.106622

68. G. Heiser, The seL4 microkernel. An introduction (2021). https://sel4.systems/About/seL4-whitepaper.pdf

69. HENSOLDT Cyber GmbH, Press release: HENSOLDT Cyber presents MiG-V, the first RISC-V processor "Made in Germany" for security applications (2020). https://hensoldt-cyber.com/wp-content/uploads/2020/05/20200515-HENSOLDT-Cyber-PM-MiG-V-is-ready-1.pdf

70. HENSOLDT Cyber GmbH, Made in Germany RISC-V (MiG-V) (2021). https://hensoldt-cyber.com/mig-v/

71. HENSOLDT Cyber GmbH, Trusted entity operating system (2021). https://hensoldt-cyber.com/trentos/

72. A. Hepp, G. Sigl, Tapeout of a RISC-V crypto chip with hardware Trojans: a case-study on Trojan design and pre-silicon detectability, in *Proceedings of the 18th ACM International Conference on Computing Frontiers, CF '21* (Association for Computing Machinery, New York, 2021), pp. 213–220. https://doi.org/10.1145/3457388.3458869

73. T. Hoque, J. Cruz, P. Chakraborty, S. Bhunia, Hardware IP trust validation: learn (the untrustworthy), and verify, in *2018 IEEE International Test Conference (ITC)* (2018), pp. 1–10. https://doi.org/10.1109/TEST.2018.8624727

74. W. Hu, L. Zhang, A. Ardeshiricham, J. Blackstone, B. Hou, Y. Tai, R. Kastner, Why you should care about don't cares: exploiting internal don't care conditions for hardware trojans, in *2017 IEEE/ACM International Conference on Computer-Aided Design (ICCAD)*, pp. 707–713 (2017). https://doi.org/10.1109/ICCAD.2017.8203846

75. Y. Hu, V.V. Menon, A. Schmidt, J. Monson, M. French, P. Nuzzo, Security-driven metrics and models for efficient evaluation of logic encryption schemes, in *Proceedings of the 17th ACM-IEEE International Conference on Formal Methods and Models for System Design, MEMOCODE '19* (Association for Computing Machinery, New York, 2019). https://doi.org/10.1145/3359986.3361207

76. S.H. Hussein, Refining a quantitative information flow metric, in *2012 5th International Conference on New Technologies, Mobility and Security (NTMS)* (IEEE, Piscataway, 2012). https://doi.org/10.1109/ntms.2012.6208689

77. IEEE graphic symbols for logic functions (includes IEEE std 91a-1991 supplement, and IEEE std 91-1984). IEEE Std 91a-1991 IEEE Std 91-1984 (1991). https://doi.org/10.1109/IEEESTD.1991.81068

78. S.A. Islam, L.K. Sah, S. Katkoori, DLockout: a design lockout technique for key obfuscated RTL IP designs, in *2019 IEEE International Symposium on Smart Electronic Systems (iSES) (Formerly iNiS)* (2019), pp. 17–20. https://doi.org/10.1109/iSES47678.2019.00017

79. S.A. Islam, L.K. Sah, S. Katkoori, High-level synthesis of key-obfuscated RTL IP with design lockout and camouflaging. ACM Trans. Des. Autom. Electron. Syst. **26**(1) (2020). https://doi.org/10.1145/3410337

80. A. Jain, U. Guin, M.T. Rahman, N. Asadizanjani, D. Duvalsaint, R.D.S. Blanton, Special session: novel attacks on logic-locking, in *2020 IEEE 38th VLSI Test Symposium (VTS)* (2020), pp. 1–10. https://doi.org/10.1109/VTS48691.2020.9107641

81. A. Jain, M.T. Rahman, U. Guin, ATPG-guided fault injection attacks on logic locking, in *2020 IEEE Physical Assurance and Inspection of Electronics (PAINE)* (2020), pp. 1–6. https://doi.org/10.1109/PAINE49178.2020.9337734

82. A. Jain, Z. Zhou, U. Guin, TAAL: tampering attack on any key-based logic locked circuits. ACM Trans. Design Autom. Electron. Syst. **26**(4), 1–22 (2021). https://doi.org/10.1145/3442379

83. R. Jarvis, M. McIntyre, Split manufacturing method for advanced semiconductor circuits. U.S. Patent no 20040102019A1, 2004

84. H.M. Kamali, K.Z. Azar, F. Farahmandi, M. Tehranipoor, Advances in logic locking: past, present, and prospects. Cryptology ePrint Archive (2022)

85. H.M. Kamali, K.Z. Azar, H. Homayoun, A. Sasan, Full-lock: hard distributions of SAT instances for obfuscating circuits using fully configurable logic and routing blocks, in *2019 56th ACM/IEEE Design Automation Conference (DAC)* (2019), pp. 1–6

86. H.M. Kamali, K.Z. Azar, H. Homayoun, A. Sasan, InterLock: an intercorrelated logic and routing locking, in *Proceedings of the 39th International Conference on Computer-Aided Design, ICCAD '20* (Association for Computing Machinery, New York, 2020). https://doi.org/10.1145/3400302.3415667

87. C. Karfa, R. Chouksey, C. Pilato, S. Garg, R. Karri, Is register transfer level locking secure? in *2020 Design, Automation & Test in Europe Conference Exhibition (DATE)* (2020), pp. 550–555. https://doi.org/10.23919/DATE48585.2020.9116261

88. R. Karmakar, S. Chattopadhyay, A particle swarm optimization guided approximate key search attack on logic locking in the absence of scan access, in *2020 Design, Automation and Test in Europe Conference Exhibition (DATE)* (2020), pp. 448–453. https://doi.org/10.23919/DATE48585.2020.9116259

89. N. Karousos, K. Pexaras, I.G. Karybali, E. Kalligeros, Weighted logic locking: a new approach for IC piracy protection, in *2017 IEEE 23rd International Symposium on On-Line Testing and Robust System Design (IOLTS)* (2017), pp. 221–226. https://doi.org/10.1109/IOLTS.2017.8046226

90. R. Karri, J. Rajendran, K. Rosenfeld, M. Tehranipoor, Trustworthy hardware: identifying and classifying hardware trojans. Computer **43**(10), 39–46 (2010). https://doi.org/10.1109/MC.2010.299

91. A. Kaur, S. Saha, C. Karfa, D. Mukhopadhyay, Corruption exposes you: statistical key recovery from compound logic locking, in *2022 23rd International Symposium on Quality Electronic Design (ISQED)*, pp. 1–6 (2022). https://doi.org/10.1109/ISQED54688.2022.9806219

92. Y. Kim, R. Daly, J. Kim, C. Fallin, J.H. Lee, D. Lee, C. Wilkerson, K. Lai, O. Mutlu, Flipping bits in memory without accessing them: an experimental study of dram disturbance errors, in *2014 ACM/IEEE 41st International Symposium on Computer Architecture (ISCA)* (2014), pp. 361–372. https://doi.org/10.1109/ISCA.2014.6853210

93. P. Kocher, J. Horn, A. Fogh, D. Genkin, D. Gruss, W. Haas, M. Hamburg, M. Lipp, S. Mangard, T. Prescher, M. Schwarz, Y. Yarom, Spectre attacks: exploiting speculative execution, in *2019 IEEE Symposium on Security and Privacy (SP)* (2019), pp. 1–19. https://doi.org/10.1109/SP.2019.00002

94. F. Koushanfar, Integrated circuits metering for piracy protection and digital rights management: an overview, in *Proceedings of the 21st Edition of the Great Lakes Symposium on Great Lakes Symposium on VLSI, GLSVLSI '11* (Association for Computing Machinery, New York, 2011), pp. 449–454. https://doi.org/10.1145/1973009.1973110

95. F. Koushanfar, G. Qu, Hardware metering, in *Proceedings of the 38th Annual Design Automation Conference, DAC '01* (Association for Computing Machinery, New York, 2001), pp. 490–493. https://doi.org/10.1145/378239.378568

96. Y. LeCun, Y. Bengio, *Convolutional Networks for Images, Speech, and Time Series* (MIT Press, Cambridge, MA, 1998), pp. 255–258

97. J. Lee, M. Tebranipoor, J. Plusquellic, A low-cost solution for protecting IPs against scan-based side-channel attacks, in *24th IEEE VLSI Test Symposium* (2006), pp. 6–99. https://doi.org/10.1109/VTS.2006.7

98. Y.W. Lee, N.A. Touba, Improving logic obfuscation via logic cone analysis, in *2015 16th Latin-American Test Symposium (LATS)* (2015), pp. 1–6. https://doi.org/10.1109/LATW.2015.7102410

99. L. Li, A. Orailoglu, Piercing logic locking keys through redundancy identification, in *2019 Design, Automation & Test in Europe Conference Exhibition (DATE)* (2019), pp. 540–545. https://doi.org/10.23919/DATE.2019.8714955

100. L. Li, A. Orailoglu, Shielding logic locking from redundancy attacks, in *2019 IEEE 37th VLSI Test Symposium (VTS)* (2019), pp. 1–6. https://doi.org/10.1109/VTS.2019.8758671

101. L. Li, A. Orailoglu, Redundancy attack: breaking logic locking through oracle-less rationality analysis. IEEE Trans. Comput.-Aided Design Integr. Circuits Syst. (2022). https://doi.org/10.1109/TCAD.2022.3192793

102. M. Li, K. Shamsi, T. Meade, Z. Zhao, B. Yu, Y. Jin, D.Z. Pan, Provably secure camouflaging strategy for IC protection. IEEE Trans. Comput.-Aided Design Integr. Circuits Syst. **38**(8), 1399–1412 (2019). https://doi.org/10.1109/TCAD.2017.2750088

103. Z. Li, F. Liu, W. Yang, S. Peng, J. Zhou, A survey of convolutional neural networks: analysis, applications, and prospects. IEEE Trans. Neural Netw. Learn. Syst. 1–21 (2021). https://doi.org/10.1109/TNNLS.2021.3084827

104. N. Limaye, E. Kalligeros, N. Karousos, I.G. Karybali, O. Sinanoglu, Thwarting all logic locking attacks: dishonest oracle with truly random logic locking. IEEE Trans. Comput.-Aided Design Integr. Circuits Syst. (2020). https://doi.org/10.1109/TCAD.2020.3029133

105. N. Limaye, S. Patnaik, O. Sinanoglu, Fa-SAT: fault-aided SAT-based attack on compound logic locking techniques, in *2021 Design, Automation & Test in Europe Conference & Exhibition (DATE)* (2021), pp. 1166–1171. https://doi.org/10.23919/DATE51398.2021.9474118

106. M. Lipp, M. Schwarz, D. Gruss, T. Prescher, W. Haas, A. Fogh, J. Horn, S. Mangard, P. Kocher, D. Genkin, Y. Yarom, M. Hamburg, Meltdown: reading kernel memory from user space, in *27th USENIX Security Symposium (USENIX Security 18)* (USENIX Association, Baltimore, 2018), pp. 973–990. https://www.usenix.org/conference/usenixsecurity18/presentation/lipp

107. Y. Liu, M. Zuzak, Y. Xie, A. Chakraborty, A. Srivastava, Strong Anti-SAT: secure and effective logic locking, in *2020 21st International Symposium on Quality Electronic Design (ISQED)* (2020), pp. 199–205. https://doi.org/10.1109/ISQED48828.2020.9136983

108. Y. Liu, M. Zuzak, Y. Xie, A. Chakraborty, A. Srivastava, Robust and attack resilient logic locking with a high application-level impact. J. Emerg. Technol. Comput. Syst. **17**(3) (2021). https://doi.org/10.1145/3446215

109. K. Lofstrom, W.R. Daasch, D. Taylor, IC identification circuit using device mismatch, in *2000 IEEE International Solid-State Circuits Conference. Digest of Technical Papers (Cat. No.00CH37056)* (2000), pp. 372–373. https://doi.org/10.1109/ISSCC.2000.839821

110. H. Mardani Kamali, K. Zamiri Azar, K. Gaj, H. Homayoun, A. Sasan, LUT-Lock: a novel LUT-based logic obfuscation for FPGA-bitstream and ASIC-hardware protection, in *2018 IEEE Computer Society Annual Symposium on VLSI (ISVLSI)* (2018), pp. 405–410. https://doi.org/10.1109/ISVLSI.2018.00080

111. J. Markoff, Old trick threatens the newest weapons. The New York Times nytimes.com (2009). www.nytimes.com/2009/10/27/science/27trojan.html

112. M.E. Massad, J. Zhang, S. Garg, M.V. Tripunitara, Logic locking for secure outsourced chip fabrication: a new attack and provably secure defense mechanism (2017). https://arxiv.org/abs/1703.10187

113. T. Meade, Z. Zhao, S. Zhang, D. Pan, Y. Jin, Revisit sequential logic obfuscation: attacks and defenses, in *2017 IEEE International Symposium on Circuits and Systems (ISCAS)* (2017), pp. 1–4. https://doi.org/10.1109/ISCAS.2017.8050606

114. S. Mitra, H.S.P. Wong, S. Wong, The Trojan-proof chip. IEEE Spectr. **52**(2), 46–51 (2015). https://doi.org/10.1109/MSPEC.2015.7024511

115. E. Moussavi, D. Sisejkovic, F. Brings, D. Kizatov, A. Singh, X.T. Vu, S. Ingebrandt, R. Leupers, V. Pachauri, F. Merchant, pHGen: a pH-based key generation mechanism using ISFETs (2022). https://doi.org/10.48550/ARXIV.2202.12085

116. E. Moussavi, D. Sisejkovic, A. Singh, D. Kizatov, R. Leupers, S. Ingebrandt, V. Pachauri, F. Merchant, A temperature independent readout circuit for ISFET-based sensor applications (2022)

117. S.A. Mário, K. Chatzikokolakis, A. McIver, C. Morgan, C. Palamidessi, G. Smith, The science of quantitative information flow, in *Information Security and Cryptography, 2020* (2020)

118. O. Mutlu, J.S. Kim, RowHammer: a retrospective. Trans. Comput.-Aided Des. Integr. Circuits Syst. **39**(8), 1555–1571 (2020). https://doi.org/10.1109/TCAD.2019.2915318

119. S. Patnaik, M. Ashraf, J. Knechtel, O. Sinanoglu, Obfuscating the interconnects: low-cost and resilient full-chip layout camouflaging, in *Proceedings of the 36th International Conference on Computer-Aided Design, ICCAD '17* (IEEE Press, 2017), pp. 41–48

120. T. Perez, S. Pagliarini, Hardware trojan insertion in finalized layouts: a silicon demonstration (2021). https://doi.org/10.48550/ARXIV.2112.02972

121. C. Pilato, A.B. Chowdhury, D. Sciuto, S. Garg, R. Karri, ASSURE: RTL locking against an untrusted foundry. IEEE Trans. Very Large Scale Integr. Syst. 1–13 (2021). https://doi.org/10.1109/TVLSI.2021.3074004

122. C. Pilato, L. Collini, L. Cassano, D. Sciuto, S. Garg, R. Karri, Optimizing the use of behavioral locking for high-level synthesis. IEEE Trans. Comput.-Aided Design Integr. Circuits Syst. (2022). https://doi.org/10.1109/TCAD.2022.3179651

123. C. Pilato, F. Regazzoni, R. Karri, S. Garg, TAO: techniques for algorithm-level obfuscation during high-level synthesis, in *2018 55th ACM/ESDA/IEEE Design Automation Conference (DAC)* (2018), pp. 1–6. https://doi.org/10.1109/DAC.2018.8465830

124. S.M. Plaza, I.L. Markov, Solving the third-shift problem in IC piracy with test-aware logic locking. IEEE Trans. Comput.-Aided Design Integr. Circuits Syst. **34**(6), 961–971 (2015). https://doi.org/10.1109/TCAD.2015.2404876

125. S.E. Quadir, J. Chen, D. Forte, N. Asadizanjani, S. Shahbazmohamadi, L. Wang, J. Chandy, M. Tehranipoor, A survey on chip to system reverse engineering. J. Emerg. Technol. Comput. Syst. **13**(1) (2016). https://doi.org/10.1145/2755563

126. M.T. Rahman, M.S. Rahman, H. Wang, S. Tajik, W. Khalil, F. Farahmandi, D. Forte, N. Asadizanjani, M. Tehranipoor, M.: defense-in-depth: a recipe for logic locking to prevail. Integr. VLSI J. **72**(C), 39–57 (2020). https://doi.org/10.1016/j.vlsi.2019.12.007

127. M.T. Rahman, S. Tajik, M.S. Rahman, M. Tehranipoor, N. Asadizanjani, The key is left under the mat: on the inappropriate security assumption of logic locking schemes, in *2020 IEEE International Symposium on Hardware Oriented Security and Trust (HOST)* (2020), pp. 262–272. https://doi.org/10.1109/HOST45689.2020.9300258

128. J. Rajendran, R. Karri, J.B. Wendt, M. Potkonjak, N. McDonald, G.S. Rose, B. Wysocki, Nano meets security: exploring nanoelectronic devices for security applications. Proc. IEEE **103**(5), 829–849 (2015). https://doi.org/10.1109/JPROC.2014.2387353

129. J. Rajendran, Y. Pino, O. Sinanoglu, R. Karri, Security analysis of logic obfuscation, in *DAC Design Automation Conference 2012* (2012), pp. 83–89. https://doi.org/10.1145/2228360.2228377

130. J. Rajendran, M. Sam, O. Sinanoglu, R. Karri, Security analysis of integrated circuit camouflaging, in *Proceedings of the 2013 ACM SIGSAC Conference on Compute and Communications Security, CCS '13* (Association for Computing Machinery, New York, 2013), pp. 709–720. https://doi.org/10.1145/2508859.2516656

131. J. Rajendran, O. Sinanoglu, R. Karri, Is split manufacturing secure? in *2013 Design, Automation & Test in Europe Conference Exhibition (DATE)* (2013), pp. 1259–1264. https://doi.org/10.7873/DATE.2013.261

132. J. Rajendran, H. Zhang, C. Zhang, G.S. Rose, Y. Pino, O. Sinanoglu, R. Karri, Fault analysis-based logic encryption. IEEE Trans. Comput. **64**(2), 410–424 (2015). https://doi.org/10.1109/TC.2013.193

133. N. Rangarajan, S. Patnaik, J. Knechtel, R. Karri, O. Sinanoglu, S. Rakheja, Opening the doors to dynamic camouflaging: harnessing the power of polymorphic devices. IEEE Trans. Emer. Topics Comput. (2020). https://doi.org/10.1109/TETC.2020.2991134

134. F. Regazzoni, S. Bhasin, A.A. Pour, I. Alshaer, F. Aydin, A. Aysu, V. Beroulle, G. Di Natale, P. Franzon, D. Hely, N. Homma, A. Ito, D. Jap, P. Kashyap, I. Polian, S. Potluri, R. Ueno, E.I. Vatajelu, V. Yli-Mäyry, Machine learning and hardware security: challenges and opportunities-invited talk, in *2020 IEEE/ACM International Conference On Computer Aided Design (ICCAD)* (2020), pp. 1–6

135. L.M. Reimann, L. Hanel, D. Sisejkovic, F. Merchant, R. Leupers, Qflow: quantitative information flow for security-aware hardware design in verilog, in *2021 IEEE 39th International Conference on Computer Design (ICCD)* (2021), pp. 603–607. https://doi.org/10.1109/ICCD53106.2021.00097

136. M.G. Rekoff, On reverse engineering. IEEE Trans. Syst. Man Cybern. **SMC-15**(2), 244–252 (1985). https://doi.org/10.1109/TSMC.1985.6313354

137. A. Rezaei, Y. Shen, H. Zhou, Rescuing logic encryption in post-SAT era by locking obfuscation, in *2020 Design, Automation & Test in Europe Conference Exhibition (DATE)* (2020), pp. 13–18. https://doi.org/10.23919/DATE48585.2020.9116500

138. S. Roshanisefat, H. Mardani Kamali, A. Sasan, SRCLock: SAT-resistant cyclic logic locking for protecting the hardware, in *Proceedings of the 2018 on Great Lakes Symposium on VLSI, GLSVLSI '18* (Association for Computing Machinery, New York, 2018), pp. 153–158. https://doi.org/10.1145/3194554.3194596

139. M. Rostami, F. Koushanfar, R. Karri, A primer on hardware security: models, methods, and metrics. Proc. IEEE **102**(8), 1283–1295 (2014). https://doi.org/10.1109/JPROC.2014.2335155

140. J.A. Roy, F. Koushanfar, I.L. Markov, EPIC: ending piracy of integrated circuits, in *2008 Design, Automation and Test in Europe* (2008), pp. 1069–1074. https://doi.org/10.1109/DATE.2008.4484823

141. J.A. Roy, F. Koushanfar, I.L. Markov, Ending piracy of integrated circuits. Computer **43**(10), 30–38 (2010). https://doi.org/10.1109/MC.2010.284

142. A. Saha, U. Chatterjee, D. Mukhopadhyay, R.S. Chakraborty, DIP learning on CAS-lock: using distinguishing input patterns for attacking logic locking, in *2022 Design, Automation and Test in Europe Conference & Exhibition (DATE)* (2022), pp. 688–693. https://doi.org/10.23919/DATE54114.2022.9774691

143. H. Salmani, M. Tehranipoor, R. Karri, On design vulnerability analysis and trust benchmarks development, in *2013 IEEE 31st International Conference on Computer Design (ICCD)* (2013), pp. 471–474. https://doi.org/10.1109/ICCD.2013.6657085

144. Semiconductor Industry Association Anti-Counterfeiting Task Force, Winning the battle against counterfeit semiconductor products (2021). https://www.semiconductors.org/wp-content/uploads/2018/01/SIA-Anti-Counterfeiting-Whitepaper.pdf

145. A. Sengupta, N. Limaye, O. Sinanoglu Breaking CAS-lock and its variants by exploiting structural traces. Cryptology ePrint Archive, Report 2021/581 (2021). https://eprint.iacr.org/2021/581

146. A. Sengupta, M. Nabeel, N. Limaye, M. Ashraf, O. Sinanoglu, Truly stripping functionality for logic locking: a fault-based perspective. IEEE Trans. Comput.-Aided Design Integr. Circuits Syst. **39**(12), 4439–4452 (2020). https://doi.org/10.1109/TCAD.2020.2968898

147. A. Sengupta, M. Nabeel, M. Yasin, O. Sinanoglu, ATPG-based cost-effective, secure logic locking, in *2018 IEEE 36th VLSI Test Symposium (VTS)* (2018), pp. 1–6. https://doi.org/10.1109/VTS.2018.8368625

148. B. Shakya, T. He, H. Salmani, D. Forte, S. Bhunia, M. Tehranipoor, Benchmarking of hardware trojans and maliciously affected circuits. J. Hardw. Syst. Secur. **1**(1), 85–102 (2017). https://doi.org/10.1007/s41635-017-0001-6

149. B. Shakya, H. Shen, M. Tehranipoor, D. Forte, Covert gates: protecting integrated circuits with undetectable camouflaging. IACR Trans. Cryptogr. Hardw. Embedded Syst. **2019**(3), 86–118 (2019). https://doi.org/10.13154/tches.v2019.i3.86-118

150. B. Shakya, X. Xu, M. Tehranipoor, D. Forte, CAS-Lock: a security-corruptibility trade-off resilient logic locking scheme. IACR Trans. Cryptogr. Hardw. Embedded Syst. **2020**(1), 175–202 (2019). https://doi.org/10.13154/tches.v2020.i1.175-202

151. K. Shamsi, M. Li, T. Meade, Z. Zhao, D.Z. Pan, Y. Jin, AppSAT: approximately deobfuscating integrated circuits, in *2017 IEEE International Symposium on Hardware Oriented Security and Trust (HOST)* (2017), pp. 95–100. https://doi.org/10.1109/HST.2017.7951805

152. K. Shamsi, M. Li, T. Meade, Z. Zhao, D.Z. Pan, Y. Jin, Cyclic obfuscation for creating sat-unresolvable circuits, in *Proceedings of the on Great Lakes Symposium on VLSI 2017,*

GLSVLSI '17 (Association for Computing Machinery, New York, 2017), pp. 173–178. https://doi.org/10.1145/3060403.3060458

153. K. Shamsi, M. Li, D.Z. Pan, Y. Jin, Cross-Lock: dense layout-level interconnect locking using cross-bar architectures, in *Proceedings of the 2018 on Great Lakes Symposium on VLSI, GLSVLSI '18* (Association for Computing Machinery, New York, 2018), pp. 147–152. https://doi.org/10.1145/3194554.3194580

154. K. Shamsi, M. Li, K. Plaks, S. Fazzari, D.Z. Pan, Y. Jin, IP protection and supply chain security through logic obfuscation: a systematic overview. ACM Trans. Des. Autom. Electron. Syst. **24**(6) (2019). https://doi.org/10.1145/3342099

155. K. Shamsi, D.Z. Pan, Y. Jin, IcySAT: improved SAT-based attacks on cyclic locked circuits, in *2019 IEEE/ACM International Conference on Computer-Aided Design (ICCAD)* (2019), pp. 1–7. https://doi.org/10.1109/ICCAD45719.2019.8942049

156. K. Shamsi, D.Z. Pan, Y. Jin, On the impossibility of approximation-resilient circuit locking, in *2019 IEEE International Symposium on Hardware Oriented Security and Trust (HOST)* (2019), pp. 161–170. https://doi.org/10.1109/HST.2019.8741035

157. C.E. Shannon, Communication theory of secrecy systems. Bell Syst. Tech. J. **28**(4), 656–715 (1949)

158. Y. Shen, Y. Li, A. Rezaei, S. Kong, D. Dlott, H. Zhou, BeSAT: behavioral SAT-based attack on cyclic logic encryption, in *Proceedings of the 24th Asia and South Pacific Design Automation Conference, ASPDAC '19* (Association for Computing Machinery, New York, 2019), pp. 657–662. https://doi.org/10.1145/3287624.3287670

159. Y. Shen, A. Rezaei, H. Zhou, SAT-based bit-flipping attack on logic encryptions, in *2018 Design, Automation & Test in Europe Conference Exhibition (DATE)* (2018), pp. 629–632. https://doi.org/10.23919/DATE.2018.8342086

160. Y. Shen, H. Zhou, Double DIP: re-evaluating security of logic encryption algorithms, in *Proceedings of the on Great Lakes Symposium on VLSI 2017, GLSVLSI '17* (Association for Computing Machinery, New York, 2017), pp. 179–184. https://doi.org/10.1145/3060403.3060469

161. Siemens EDA Software, ModelSim. https://eda.sw.siemens.com/en-US/ic/modelsim/

162. S. Sirone, P. Subramanyan, Functional analysis attacks on logic locking, in *2019 Design, Automation and Test in Europe Conference Exhibition (DATE)* (2019), pp. 936–939. https://doi.org/10.23919/DATE.2019.8715163

163. S. Sirone, P. Subramanyan, Functional analysis attacks on logic locking. IEEE Trans. Inform. Forensics Secur. **15**, 2514–2527 (2020). https://doi.org/10.1109/TIFS.2020.2968183

164. D. Sisejkovic, Designing trustworthy hardware with logic locking. Dissertation, Rheinisch-Westfälische Technische Hochschule Aachen, Aachen (2022). https://doi.org/10.18154/RWTH-2022-02625

165. D. Sisejkovic, L. Collini, B. Tan, C. Pilato, R. Karri, R. Leupers, Designing ML-resilient locking at register-transfer level (2022). https://doi.org/10.48550/ARXIV.2203.05399

166. D. Sisejkovic, F. Merchant, L.M. Reimann, R. Leupers, Deceptive logic locking for hardware integrity protection against machine learning attacks. IEEE Trans. Comput.-Aided Design Integr. Circuits Syst. (2021). https://doi.org/10.1109/TCAD.2021.3100275

167. D. Šišejković, F. Merchant, L.M. Reimann, R. Leupers, S. Kegreiß, Scaling logic locking schemes to multi-module hardware designs, in *Architecture of Computing Systems—ARCS 2020*, ed. by A. Brinkmann, W. Karl, S. Lankes, S. Tomforde, T. Pionteck, C. Trinitis (Springer, Cham, 2020), pp. 138–152

168. D. Sisejkovic, F. Merchant, L.M. Reimann, H. Srivastava, A. Hallawa, R. Leupers, Challenging the security of logic locking schemes in the era of deep learning: a neuroevolutionary approach (2020). https://arxiv.org/abs/2011.10389

169. D. Sisejkovic, F. Merchant, L.M. Reimann, H. Srivastava, A. Hallawa, R. Leupers, Challenging the security of logic locking schemes in the era of deep learning: a neuroevolutionary approach. J. Emerg. Technol. Comput. Syst. **17**(3) (2021). https://doi.org/10.1145/3431389

170. D. Sisejkovic, L.M. Reimann, E. Moussavi, F. Merchant, R. Leupers, Logic locking at the frontiers of machine learning: a survey on developments and opportunities, in *2021*

IFIP/IEEE 29th International Conference on Very Large Scale Integration (VLSI-SoC) (2021), pp. 1–6. https://doi.org/10.1109/VLSI-SoC53125.2021.9606979

171. M. Soos, K. Nohl, C. Castelluccia, Extending SAT solvers to cryptographic problems, in *Theory and Applications of Satisfiability Testing—SAT 2009*, ed. by O. Kullmann (Springer, Berlin, Heidelberg, 2009), pp. 244–257

172. F. Staudigl, H.A. Indari, D. Schoen, D. Sisejkovic, F. Merchant, J.M. Joseph, V. Rana, S. Menzel, R. Leupers, NeuroHammer: inducing bit-flips in memristive crossbar memories, in *2022 Design, Automation & Test in Europe Conference & Exhibition (DATE)* (2022), pp. 1181–1184. https://doi.org/10.23919/DATE54114.2022.9774651

173. F. Staudigl, F. Merchant, R. Leupers, A survey of neuromorphic computing-in-memory: architectures, simulators, and security. IEEE Design Test **39**(2), 90–99 (2022). https://doi.org/10.1109/MDAT.2021.3102013

174. F. Staudigl, K.J.X. Sturm, M. Bartel, T. Fetz, D. Sisejkovic, J.M. Joseph, L.B. Pöhls, R. Leupers, X-fault: impact of faults on binary neural networks in memristor-crossbar arrays with logic-in-memory computation (2022). https://doi.org/10.48550/ARXIV.2204.01501

175. P. Subramanyan, S. Ray, S. Malik, Evaluating the security of logic encryption algorithms, in *2015 IEEE International Symposium on Hardware Oriented Security and Trust (HOST)* (2015), pp. 137–143. https://doi.org/10.1109/HST.2015.7140252

176. P. Subramanyan, N. Tsiskaridze, W. Li, A. Gascón, W.Y. Tan, A. Tiwari, N. Shankar, S.A. Seshia, S. Malik, Reverse engineering digital circuits using structural and functional analyses. IEEE Trans. Emerg. Topics Comput. **2**(1), 63–80 (2014). https://doi.org/10.1109/TETC.2013.2294918

177. J. Sweeney, M.J.H. Heule, L. Pileggi, Modeling techniques for logic locking, in *2020 IEEE/ACM International Conference On Computer Aided Design (ICCAD)* (2020), pp. 1–9

178. Synopsys Inc., ASIP designer. https://www.synopsys.com/dw/ipdir.php?ds=asip-designer

179. Synopsys Inc., Design compiler. https://www.synopsys.com/implementation-and-signoff/rtl-synthesis-test/dc-ultra.html

180. Synopsys Inc., Formality. https://www.synopsys.com/verification.html

181. C. Szegedy, S. Ioffe, V. Vanhoucke, A.A. Alemi, Inception-v4, Inception-ResNet and the impact of residual connections on learning, in *Proceedings of the Thirty-First AAAI Conference on Artificial Intelligence, AAAI'17* (AAAI Press, 2017), pp. 4278–4284

182. S. Takamaeda-Yamazaki, Pyverilog: a python-based hardware design processing toolkit for verilog HDL, in *Applied Reconfigurable Computing*, ed. by K. Sano, D. Soudris, M. Hübner, P.C. Diniz (Springer, Cham, 2015), pp. 451–460

183. B. Tan, R. Karri, Challenges and new directions for ai and hardware security, in *2020 IEEE 63rd International Midwest Symposium on Circuits and Systems (MWSCAS)* (2020), pp. 277–280. https://doi.org/10.1109/MWSCAS48704.2020.9184612

184. F. Tehranipoor, N. Karimian, M.M. Kermani, H. Mahmoodi, Deep RNN-oriented paradigm shift through BOCANet: broken obfuscated circuit attack (2018). https://arxiv.org/abs/1803.03332

185. F. Tehranipoor, N. Karimian, M. Mozaffari Kermani, H. Mahmoodi, Deep RNN-oriented paradigm shift through BOCANet: broken obfuscated circuit attack, in *Proceedings of the 2019 on Great Lakes Symposium on VLSI, GLSVLSI '19* (Association for Computing Machinery, New York, 2019), pp. 335–338. https://doi.org/10.1145/3299874.3318031

186. M. Tehranipoor, F. Koushanfar, A survey of hardware Trojan taxonomy and detection. IEEE Design Test Comput. **27**(1), 10–25 (2010). https://doi.org/10.1109/MDT.2010.7

187. R. Torrance, D. James, The state-of-the-art in semiconductor reverse engineering, in *Proceedings of the 48th Design Automation Conference, DAC '11* (Association for Computing Machinery, New York, 2011), pp. 333–338. https://doi.org/10.1145/2024724.2024805

188. A. Traber, M. Gautschi, P.D. Schiavone, RI5CY: user manual. https://pulp-platform.org/docs/ri5cy_user_manual.pdf

189. UC Berkeley Architecture Research, RISC-V torture test generator. https://github.com/ucb-bar/riscv-torture

190. I. Verbauwhede, AES and other secret key implementations (2011). https://www.cosic.esat.
kuleuven.be/ecrypt/courses/albena11/slides/ingrid_verbauwhede_aes_implementations.pdf
191. Verified Market Research, Semiconductor IP market size and forecast (2021). https://www.
verifiedmarketresearch.com/product/global-semiconductor-intellectual-property-market/
192. D. Šišejković, R. Leupers, G. Ascheid, S. Metzner, A unifying logic encryption security
metric, in *Proceedings of the 18th International Conference on Embedded Computer Systems:
Architectures, Modeling, and Simulation, SAMOS '18* (Association for Computing Machinery, New York, 2018), pp. 179–186. https://doi.org/10.1145/3229631.3229636
193. D. Šišejković, F. Merchant, R. Leupers, Protecting the integrity of processor cores with logic
encryption, in *2019 32nd IEEE International System-on-Chip Conference (SOCC)* (2019),
pp. 424–425. https://doi.org/10.1109/SOCC46988.2019.1570564157
194. D. Šišejković, F. Merchant, R. Leupers, G. Ascheid, S. Kegreiss, Control-lock: securing
processor cores against software-controlled hardware Trojans, in *Proceedings of the 2019
on Great Lakes Symposium on VLSI, GLSVLSI '19* (Association for Computing Machinery,
New York, 2019), pp. 27–32. https://doi.org/10.1145/3299874.3317983
195. D. Šišejković, F. Merchant, L.M. Reimann, R. Leupers, M. Giacometti, S. Kegreiß, A secure
hardware-software solution based on RISC-V, logic locking and microkernel, in *Proceedings
of the 23th International Workshop on Software and Compilers for Embedded Systems,
SCOPES '20* (Association for Computing Machinery, New York, 2020), pp. 62–65. https://
doi.org/10.1145/3378678.3391886
196. Y. Wang, P. Liu, X. Han, Y. Jiang, Hardware trojan detection method for inspecting integrated
circuits based on machine learning, in *2021 22nd International Symposium on Quality
Electronic Design (ISQED)* (2021), pp. 432–436. https://doi.org/10.1109/ISQED51717.2021.
9424314
197. A. Waterman, Y. Lee, D. Patterson, K. Asanovic, The RISC-V instruction set manual. Volume
I: user-level ISA, version 2.0, Tech. Rep. UCB/EECS-2014-54 (2014)
198. M. Werner, B. Lippmann, J. Baehr, H. Gräb, Reverse engineering of cryptographic cores
by structural interpretation through graph analysis, in *2018 IEEE 3rd International Verification and Security Workshop (IVSW)* (2018), pp. 13–18. https://doi.org/10.1109/IVSW.2018.
8494896
199. S. Williams, The Icarus verilog compilation system (2017). http://iverilog.icarus.com
200. C. Wolf, Yosys open synthesis suite. http://www.clifford.at/yosys/
201. K. Xiao, D. Forte, Y. Jin, R. Karri, S. Bhunia, M. Tehranipoor, Hardware Trojans: lessons
learned after one decade of research. ACM Trans. Des. Autom. Electron. Syst. **22**(1) (2016).
https://doi.org/10.1145/2906147
202. K. Xiao, D. Forte, M. Tehranipoor, A novel built-in self-authentication technique to prevent
inserting hardware Trojans. IEEE Trans. Comput.-Aided Design Integr. Circuits Syst. **33**(12),
1778–1791 (2014). https://doi.org/10.1109/TCAD.2014.2356453
203. K. Xiao, M. Tehranipoor, BISA: built-in self-authentication for preventing hardware Trojan
insertion, in *2013 IEEE International Symposium on Hardware-Oriented Security and Trust
(HOST)* (2013), pp. 45–50. https://doi.org/10.1109/HST.2013.6581564
204. Y. Xie, A. Srivastava, Mitigating SAT attack on logic locking, in *Cryptographic Hardware
and Embedded Systems—CHES 2016*, ed. by B. Gierlichs, A.Y. Poschmann (Springer, Berlin,
Heidelberg, 2016), pp. 127–146
205. Y. Xie, A. Srivastava, Delay locking: security enhancement of logic locking against IC
counterfeiting and overproduction, in *Proceedings of the 54th Annual Design Automation
Conference 2017, DAC '17* (Association for Computing Machinery, New York, 2017). https://
doi.org/10.1145/3061639.3062226
206. Y. Xie, A. Srivastava, Anti-SAT: mitigating SAT attack on logic locking. IEEE Trans.
Comput.-Aided Design Integr. Circuits Syst. **38**(2), 199–207 (2019). https://doi.org/10.1109/
TCAD.2018.2801220
207. X. Xu, B. Shakya, M.M. Tehranipoor, D. Forte, Novel bypass attack and BDD-based tradeoff
analysis against all known logic locking attacks, in *Cryptographic Hardware and Embedded
Systems—CHES 2017*. Lecture Notes in Computer Science, vol. 10529 (Springer, Berlim,
2017), pp. 189–210

208. M. Xue, C. Gu, W. Liu, S. Yu, M. O'Neill, Ten years of hardware trojans: a survey from the attacker's perspective. IET Comput. Digital Tech. **14**(6), 231–246 (2020)

209. F. Yang, M. Tang, O. Sinanoglu, Stripped functionality logic locking with hamming distance-based restore unit (SFLL-HD)—unlocked. IEEE Trans. Inform. Forensics Secur. **14**(10), 2778–2786 (2019). https://doi.org/10.1109/TIFS.2019.2904838

210. S. Yang, Logic synthesis and optimization benchmarks version 3.0. Tech. Report, Microelectronics Centre of North Carolina (1991)

211. X.M. Yang, P.P. Chen, H.Y. Chiang, C.C. Lin, Y.C. Chen, C.Y. Wang, Looplock 2.0: an enhanced cyclic logic locking approach. IEEE Trans. Comput.-Aided Design Integr. Circuits Syst. **41**(1), 29–34 (2022). https://doi.org/10.1109/TCAD.2021.3053912

212. Y. Yang, Z. Chen, Y. Liu, T.Y. Ho, Y. Jin, P. Zhou, How secure is split manufacturing in preventing hardware Trojan? ACM Trans. Des. Autom. Electron. Syst. **25**(2) (2020). https://doi.org/10.1145/3378163

213. R. Yasaei, S. Faezi, M.A.A. Faruque, Golden reference-free hardware trojan localization using graph convolutional network (2022). https://doi.org/10.48550/ARXIV.2207.06664

214. M. Yasin, B. Mazumdar, S.S. Ali, O. Sinanoglu, Security analysis of logic encryption against the most effective side-channel attack: DPA, in *2015 IEEE International Symposium on Defect and Fault Tolerance in VLSI and Nanotechnology Systems (DFTS)* (2015), pp. 97–102. https://doi.org/10.1109/DFT.2015.7315143

215. M. Yasin, B. Mazumdar, J.J.V. Rajendran, O. Sinanoglu, SARLock: SAT attack resistant logic locking, in, *2016 IEEE International Symposium on Hardware Oriented Security and Trust (HOST)* (2016), pp. 236–241. https://doi.org/10.1109/HST.2016.7495588

216. M. Yasin, B. Mazumdar, O. Sinanoglu, J. Rajendran, Security analysis of Anti-SAT, in *2017 22nd Asia and South Pacific Design Automation Conference (ASP-DAC)* (2017), pp. 342–347. https://doi.org/10.1109/ASPDAC.2017.7858346

217. M. Yasin, B. Mazumdar, O. Sinanoglu, J. Rajendran, Removal attacks on logic locking and camouflaging techniques. IEEE Trans. Emerg. Topics Comput. **8**(2), 517–532 (2020). https://doi.org/10.1109/TETC.2017.2740364

218. M. Yasin, J.J. Rajendran, O. Sinanoglu, *Trustworthy Hardware Design: Combinational Logic Locking Techniques* (Springer, Berlim, 2020). https://doi.org/10.1007/978-3-030-15334-2

219. M. Yasin, J.J. Rajendran, O. Sinanoglu, R. Karri, On improving the security of logic locking. IEEE Trans. Comput.-Aided Design Integr. Circuits Syst. **35**(9), 1411–1424 (2016). https://doi.org/10.1109/TCAD.2015.2511144

220. M. Yasin, S.M. Saeed, J. Rajendran, O. Sinanoglu, Activation of logic encrypted chips: pre-test or post-test? in *2016 Design, Automation & Test in Europe Conference Exhibition (DATE)* (2016), pp. 139–144

221. M. Yasin, A. Sengupta, M.T. Nabeel, M. Ashraf, J.J. Rajendran, O. Sinanoglu, Provably-secure logic locking: from theory to practice, in *Proceedings of the 2017 ACM SIGSAC Conference on Computer and Communications Security, CCS '17* (Association for Computing Machinery, New York, 2017), pp. 1601–1618. https://doi.org/10.1145/3133956.3133985

222. M. Yasin, A. Sengupta, B.C. Schafer, Y. Makris, O. Sinanoglu, J.J. Rajendran, What to lock? functional and parametric locking, in *Proceedings of the on Great Lakes Symposium on VLSI 2017, GLSVLSI '17* (Association for Computing Machinery, New York, 2017), pp. 351–356. https://doi.org/10.1145/3060403.3060492

223. M. Yasin, O. Sinanoglu, Evolution of logic locking, in *2017 IFIP/IEEE International Conference on Very Large Scale Integration (VLSI-SoC)* (2017), pp. 1–6. https://doi.org/10.1109/VLSI-SoC.2017.8203496

224. K. Zamiri Azar, H. Mardani Kamali, H. Homayoun, A. Sasan, Threats on logic locking: a decade later, in *Proceedings of the 2019 on Great Lakes Symposium on VLSI, GLSVLSI '19* (Association for Computing Machinery, New York, 2019), pp. 471–476. https://doi.org/10.1145/3299874.3319495

225. F. Zaruba, L. Benini, The cost of application-class processing: energy and performance analysis of a Linux-ready 1.7-GHz 64-bit RISC-V core in 22-nm FDSOI technology. IEEE Trans. Very Large Scale Integr. Syst. **27**(11), 2629–2640 (2019). https://doi.org/10.1109/TVLSI.2019.2926114

226. Y. Zhang, P. Cui, Z. Zhou, U. Guin, TGA: an oracle-less and topology-guided attack on logic locking, in *Proceedings of the 3rd ACM Workshop on Attacks and Solutions in Hardware Security Workshop, ASHES'19* (Association for Computing Machinery, New York, 2019), pp. 75–83. https://doi.org/10.1145/3338508.3359576

227. H. Zhou, A humble theory and application for logic encryption. IACR Cryptol. ePrint Arch. **2017**, 696 (2017)

228. H. Zhou, R. Jiang, S. Kong, Cycsat: SAT-based attack on cyclic logic encryptions, in *2017 IEEE/ACM International Conference on Computer-Aided Design (ICCAD)* (2017), pp. 49–56. https://doi.org/10.1109/ICCAD.2017.8203759

229. H. Zhou, A. Rezaei, Y. Shen, Resolving the trilemma in logic encryption, in *2019 IEEE/ACM International Conference on Computer-Aided Design (ICCAD)* (2019), pp. 1–8. https://doi.org/10.1109/ICCAD45719.2019.8942076

230. J. Zhou, X. Zhang, Generalized sat-attack-resistant logic locking. IEEE Trans. Inform. Forensics Secur. **16**, 2581–2592 (2021). https://doi.org/10.1109/TIFS.2021.3059271

231. D. Šišejković, F. Merchant, R. Leupers, G. Ascheid, S. Kegreiss, Inter-Lock: logic encryption for processor cores beyond module boundaries, in *2019 IEEE European Test Symposium (ETS)* (2019), pp. 1–6. https://doi.org/10.1109/ETS.2019.8791528

232. D. Šišejković, F. Merchant, R. Leupers, G. Ascheid, V. Kiefer, A critical evaluation of the paradigm shift in the design of logic encryption algorithms, in *2019 International Symposium on VLSI Design, Automation and Test (VLSI-DAT)* (2019), pp. 1–4. https://doi.org/10.1109/VLSI-DAT.2019.8741531

Index

© The Author(s), under exclusive license to Springer Nature Switzerland AG 2023
D. Sisejkovic, R. Leupers, *Logic Locking*,
https://doi.org/10.1007/978-3-031-19123-7